Peterson
First Guide

Wildlife

Written and Illustrated by
Sarah Landry

HOUGHTON MIFFLIN COMPANY
Boston   New York

PETERSON FIRST GUIDES, PETERSON FIELD
GUIDES, and PETERSON FIELD GUIDE SERIES
are registered trademarks of
Houghton Mifflin Company.

For information about permission to reproduce
selections from this book, write to Permissions,
Houghton Mifflin Company, 215 Park Avenue
South, New York, New York 10003

Library of Congress Cataloging-in-Publication Data
Landry, Sarah.
    Peterson first guide to urban wildlife / text and
illustrations by Sarah Landry.
        p.   cm.
    Cover title: Peterson first guides. Urban wildlife.
    Includes index.
    ISBN 0-395-93544-X
    1. Urban fauna--North America--Identification.
I. Title. II. Title: Urban wildlife. III. Title: Peterson
first guides. Urban wildlife.
QL151.L35   1994
591.97'09173'2--dc20
                                          93-31279
                                          CIP

Printed in China

SCP 20 19 18 17 16 15 14 13 12

4500351521

Editor's Note

In 1934, my Field Guide to the Birds first saw the light of day. This book was designed so that live birds could be readily identified at a distance, by their patterns, shapes, and field marks, without resorting to the technical points specialists use to name species in the hand or in the specimen tray. The book introduced the "Peterson System," as it is now called, a visual system based on patternistic drawings with arrows to pinpoint the key field marks. The system is now used throughout the Peterson Field Guide series, which has grown to over 40 volumes on a wide range of subjects, from ferns to fishes, rocks to stars, animal tracks to edible plants.

Even though Peterson Field Guides are intended for the novice as well as the expert, there are still many beginners who would like something simpler to start with — a smaller guide that would give them confidence. It is for this audience — those who perhaps recognize a crow or a robin, buttercup or daisy, but little else — that the Peterson First Guides have been created. They offer a selection of the animals and plants you are most likely to see during your first forays afield. By narrowing the choices — and using the Peterson System — they make identification even simpler. First Guides make it easy to get started in the field, and easy to graduate to the full-fledged Peterson Field Guides. This one gives the beginner a start on the animals and plants that make their homes near, and sometimes in, ours. As Sarah Landry shows us in these pages, we don't have to go to wild places to find wildlife. A surprisingly wide range of species can be found in our cities and towns, from familiar animals like the Raccoon to more exotic ones like the Mountain Lion. The more you look, the more you will see.

Roger Tory Peterson

Introducing Urban Wildlife

By the year 2000, about 90 percent of the world's people will live in cities, towns, and suburbs. In a few centuries, our growing population and our technologies have changed the planet more than all the long 400 million years that life has been evolving on Earth.

As we take over wild land for our own uses, we dislodge and destroy many species of living things. A few hardy species, however, benefit from the changes we create. Only a tiny fraction of the estimated 10 million to 100 million life forms on Earth can survive in the severely disrupted environments we are creating. You will meet some of these adaptable plants and animals in this book.

Some of these species, like the cottontail (page 102), are native to North America, and simply like the ways we have altered the land.

Others are not native to North America but were brought here — accidentally or on purpose — by us. These introduced, or exotic, species, such as the carp (see page 56), have been successful at exploiting the niches and opportunities offered by their new homes. These species may compete with and displace native species.

4

Many people, when they think of urban wildlife, imagine squirrels and pigeons, or perhaps even foxes and peregrine falcons. It is true that compared with the great diversity of species on Earth, there are relatively few species of urban wildlife. But we have more nonhuman neighbors than we might suppose — more, in some cases, than we might think we want to know about. Like most neighbors, these living things become less foreign and more interesting once we come to know and understand them better.

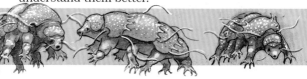

A word of warning: Many wild creatures are potential carriers of diseases and should not be touched. Raccoons, foxes, skunks, bats, rats, mice, and other mammals, as well as ticks and flies, can carry serious diseases such as rabies and Lyme disease. They can transmit these diseases to humans or to pets by bites or by contaminating food. Even an animal killed by a car can transmit disease if it is handled. Remember too that almost all wild animals are protected by conservation laws and cannot be kept as pets without a special license.

This book is divided into six sections, or Kingdoms: bacteria, viruses, protoctists, fungi, animals, and plants. The chart below shows approximately how many species are in each kingdom. As you can see, humans and other animals with backbones make up only a tiny sliver of the range of life forms on Earth.

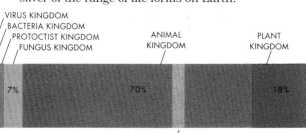

VIRUS KINGDOM
BACTERIA KINGDOM
PROTOCTIST KINGDOM
FUNGUS KINGDOM

ANIMAL KINGDOM

PLANT KINGDOM

7%   70%   18%

ANIMALS WITH BACKBONE
(INCLUDING US)

Bacteria Kingdom

Perhaps the most important organisms on Earth are the bacteria. There are only about 2,000 species, but they are everywhere in huge numbers. They have only one cell and are tiny—a chain of 1,300 bacteria would span the head of a pin. Bacteria come in three simple forms: round, rod-shaped, and spiral. The three forms are shown opposite.

It is sometimes hard to decide whether a bacterium is "friendly" or "unfriendly." Bacteria create chemicals while breaking down material to make their food, and occasionally these by-products are toxic to other organisms. A by-product of an ancient bacterium was oxygen, which was a poisonous gas then, but is vital to us now. The bacterium *Salmonella* causes food poisoning in humans. Bacteria cause diseases like tuberculosis, typhoid, and cholera. On the other hand, other bacteria make new life possible by breaking down and reassembling organic material such as dead plants and animals. This process recycles the elements inside, making them available for use by other living things.

INTESTINAL BACTERIA
Breaking down and absorbing materials we don't need, *Escherichia coli* bacteria—*E. coli*, for short—usually live in our intestines, most often harmlessly, To give a sense of their size, they are shown here lounging on the point of a pin.

NITROGEN-FIXING BACTERIA
Although the air we breathe is 78 percent nitrogen, and plants cannot grow without it, it is not available to them without the aid of these bacteria. These form associations with the roots of many plants in nodules which the plant and the bacterium jointly produce. In exchange for carbohydrates made by the plant and sunlight, the bacteria furnish the plant with nitrogen. Without nitrogen-fixing bacteria, the Earth would look like the surface of the moon.

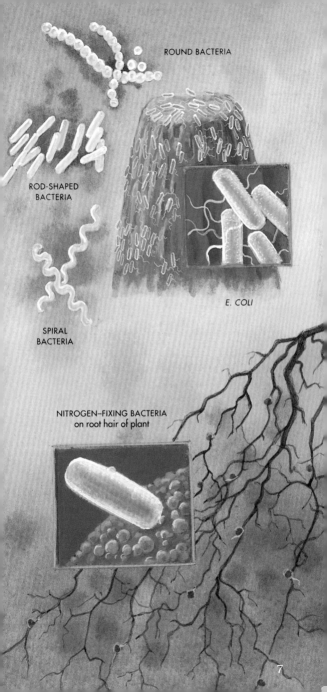

ROUND BACTERIA

ROD-SHAPED
BACTERIA

SPIRAL
BACTERIA

E. COLI

NITROGEN–FIXING BACTERIA
on root hair of plant

7

Virus Kingdom

Viruses are infinitesimally tiny, much smaller than bacteria—a chain of 10,000 common cold viruses would span the head of a pin. They are simply strands of genetic code wrapped in a coat of protein. In fact, it is debatable whether viruses qualify as living things: they cannot feed or grow; they cannot even reproduce by themselves, but instead insert their coded instructions into a host cell, forcing the host cell to make more viruses.

Once viruses penetrate cells, they cause human diseases like the common cold and the flu, mumps, rabies, polio, and AIDS.

BACTERIOPHAGES

Bacteriophages, or phages, are viruses that infect bacterial cells. Their design—a cross between a hypodermic needle and a moon lander—shows how phages work. The phages shown here are invading an *E. coli* bacterium. The phage on the left has attached itself to the wall of the bacteria and is about to inject its strand of "instructions" into the bacterium. The ghostly phages on the right have already done so. The bacterium will now make more viruses instead of more bacteria. Someday, we may be able to make viruses carry beneficial, instead of harmful, ingredients to cells.

COMMON COLD VIRUS

The virus that causes the common cold is especially tiny. Here, it has found a thin spot in the protective mucus of the throat and is about to "dock" with a human throat cell by fitting projections on the cell's surface into crevices in its own body. Then it will take over the cell's reproductive machinery, causing it to make more viruses. Now the infection has begun. You feel a sore throat. As your body's immune system mobilizes to fight the viruses, your nose runs and your temperature rises. You have a cold.

BACTERIOPHAGE

"GHOST"
BACTERIOPHAGES

BACTERIAL CELL

COMMON
COLD
VIRUS

HUMAN
THROAT
CELL

Protoctist Kingdom

We are neighbors with thousands of these microscopic, mostly one-celled organisms without realizing it; they live everywhere in water, from harbors to ponds to flower vases.

DINOFLAGELLATES

These one-celled sea creatures whirl constantly using a taillike flagellum. Some use chlorophyll to make food, as plants do. Shown here are *Noctiluca* and *Gonyaulax*, which give off flashes of brilliant light when disturbed. Where there are no competing lights in summer, our harbors glow at night with a magical, cold, blue light made by these bioluminescent organisms.

AMOEBAS

Like tiny monsters, amoebas engulf their prey with jellylike arms called pseudopods. Some cause diseases like dysentery. Shown here is *Arcella*, an amoeba common in ponds, ditches, and gutters. It lives in a protective case called a test and reaches out a pseudopod to feed. Also shown is *Entamoeba gingivalis*, which lives in your mouth, where it feeds on loose cells and bacteria.

FLAGELLATES

These organisms live mostly in fresh water, propelling themselves erratically with flagella. Like many other euglenas, *Euglena gracilis*, shown here, can either ingest its food or make its own from chlorophyll and sunlight, like a plant.

CILIATES

Ciliates can direct their movements more precisely by moving hairlike cilia. In only one cell, they have the equivalent of mouths, throats, guts, and anuses. The *Paramecium* shown here is being eaten by a predator ciliate, *Didinium*, after a battle during which both organisms used arsenals of tiny "missiles." This drama might occur in the flower vase on your table.

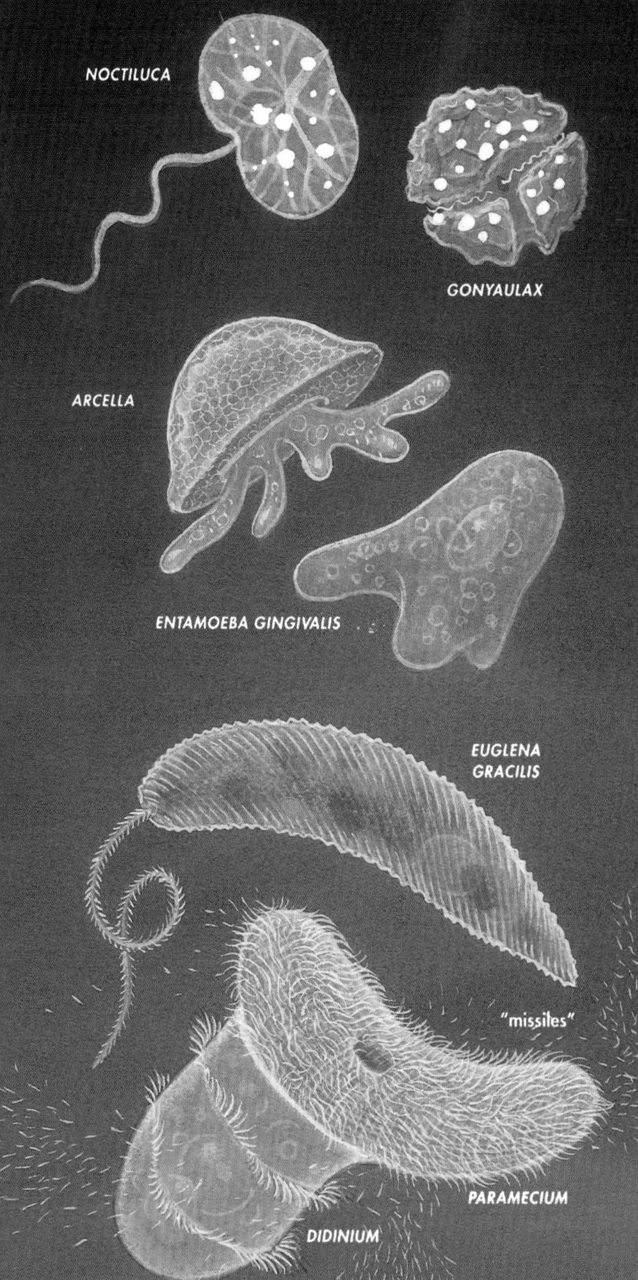

NOCTILUCA

GONYAULAX

ARCELLA

ENTAMOEBA GINGIVALIS

EUGLENA GRACILIS

"missiles"

PARAMECIUM

DIDINIUM

11

BROWN ALGAE

Known to us as brown seaweeds, these organisms are exceptions to the "one-cell" rule that applies to other protoctists. Although they are not plants, brown algae make their own food through photosynthesis, with the aid of chloroplasts and sunlight. Fronds of dead "seaweed" are often found washed up in ports and seacoast towns.

ROCKWEED   To 3 ft.

Rockweed is found along northern ocean shores. The species shown here is widely distributed. It can be seen attached by its "holdfasts" to pilings and rocks, held afloat by its *air bladders* and buoyant reproductive receptacles.

SARGASSUM WEED   2 ft. or longer

Sargassum Weed usually drifts in huge, golden masses far out in the Atlantic Ocean. Multitudes of other organisms take shelter in its *narrow fronds*. Small pieces occasionally break off and drift ashore along the southeastern and Gulf coasts.

GIANT KELP   To 200 ft.

Kelps are large species of algae that grow on the ocean floors off both North American coasts. Giant Kelp grows in dense beds off the Pacific Coast. Attached to the ocean floor by rootlike "holdfasts," the tall stems support 3-foot blades buoyed up by individual *air bladders*. Sections of this magnificent organism sometimes break off in storms and drift ashore.

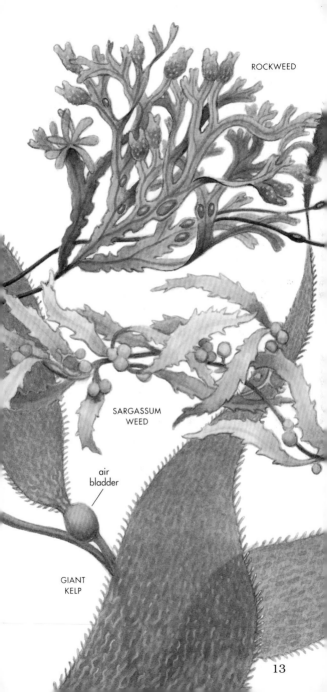

ROCKWEED

SARGASSUM
WEED

air
bladder

GIANT
KELP

13

Fungus Kingdom

Mushrooms, molds, rusts, and yeasts belong to this kingdom. A fungus consists of thread-like, almost invisible strands called hyphae. These produce the part of the fungus you can see, the "fruiting bodies" that produce the tiny reproductive spores. Like bacteria, fungi are important recyclers of dead organic matter.

BREAD MOLDS

Spores of two kinds of common bread mold, green and black, are shown here seizing the humid moment when they can colonize their host food. Even a tiny amount of bread mold has such a potent taste that we reject the bread—quite properly, from the mold's point of view.

COMMON YEAST

Bread yeast smells and tastes delicious. Its plump little cells grow by budding rather than by creating spores. Multiplying very rapidly, yeasts break down sugars in bread dough, creating carbon dioxide bubbles that cause the dough to rise.

AGARICUS MUSHROOM   3 in.

Familiar on pizzas and in salads, the fruiting body of this fungus is the common supermarket mushroom. Many other mushrooms are highly poisonous.

ATHLETE'S FOOT FUNGUS
PENICILLIN

Ironically, the athlete's foot fungus, which feeds on nutrients produced by our skin glands and releases itchy by-products, is in the same fungal group as penicillin. Penicillin fungi discharge poisons that kill bacteria competing with them for food.

SHIELD LICHEN   3 in.

Lichens are partnerships between a fungus and a green alga. Growing slowly on stones or bark, these long-lived organisms are highly sensitive to air pollution and will die or fail to grow where pollution is present.

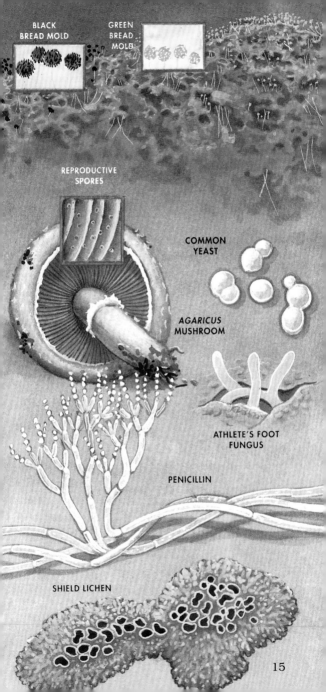

BLACK BREAD MOLD

GREEN BREAD MOLD

REPRODUCTIVE SPORES

COMMON YEAST

AGARICUS MUSHROOM

ATHLETE'S FOOT FUNGUS

PENICILLIN

SHIELD LICHEN

15

Animal Kingdom

Many-celled organisms from sponges to man belong to this huge kingdom.

SPONGES

Sponges are aquatic animals with numerous pores that channel food, wastes, and oxygen in and out of their cells. Fancy bathroom sponges are the animals' fibrous skeletons.

COELENTERATES

This aquatic group includes sea anemones, corals, and jellyfish. They have mouths surrounded by feeding tentacles armed with stinging cells.

MOON JELLYFISH   16 in.
Capable of only the simplest coordinated actions, the Moon Jellyfish, with its four loops of reproductive organs, drifts helplessly onto beaches and into harbors.

ROTIFERS

Rotifers are very common freshwater plankton found in wet places from lakes to wet mosses. They are named for the rhythmic movement of the *cilia* around their mouths.

NEMATODES

Living in the soil or parasitic in plants or animals, these unsegmented "roundworms" are everywhere. All nematode worms look much alike: they are round and pointed at both ends. Some are tiny; some are several feet long. Parasitic nematodes are a serious health problem in the tropics.

PIN WORM $1/2$ in.
Children are occasionally invaded by pin worms. They live in our guts and lay eggs at the nearest exit—the anus. The eggs cause itching, are scratched, and are transferred back to the mouth, restarting the cycle.

HEARTWORM   To 12 in.
This nematode is transferred by mosquitoes to the heart tissues of our dogs.

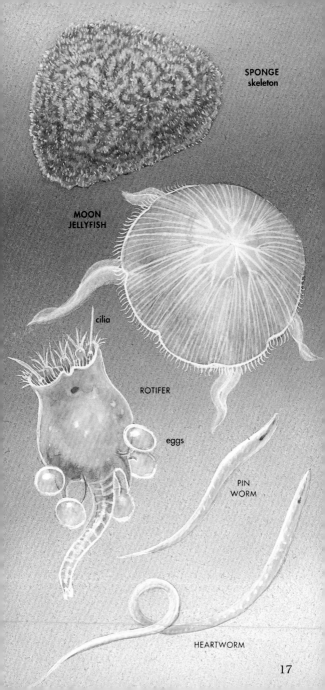

SPONGE skeleton

MOON JELLYFISH

cilia

ROTIFER

eggs

PIN WORM

HEARTWORM

17

MOLLUSKS

All the members of this large group have a hard shell, either an external one like that of the clams or an internal one like that of the 60-foot giant squid. Most have a unique, abrasive "tongue" called a radula that helps them feed, and they move about with the help of a muscular "foot." Two of the largest groups of mollusks are the gastropods (single-shelled animals such as snails) and the bivalves (two-shelled animals such as clams and oysters).

COMMON MUSSEL   4 in.

This *bluish* bivalve and its western cousin, the ribbed California Mussel (10 in.), anchor themselves with tough threads.

ZEBRA MUSSEL   3/4 in.

Brought here in the freshwater ballast of ships from Europe, the *striped* Zebra Mussel has no natural predators in North America and is spreading throughout fresh waters of the East, clogging intake pipes of public water supplies and fouling turbines

SHIPWORM   5 in.

Shipworms eat wood exposed to salt water. Their shells are modified into effective *drills* that grind tunnels in wooden pilings and ships.

PERIWINKLE   1 in.

The Periwinkle is found on pilings and rocks along all our coastlines. When out of water, it closes up and breathes trapped, moist air through its gills.

GARDEN SNAIL   3/4 in.

Introduced from Europe, the *striped* Garden Snail has a primitive lung and lives on land, slipping along on a mucus trail. A gourmet delicacy abroad, here they are regarded mainly as a garden pest.

GARDEN SLUG   2 1/2 in.

Like snails, slugs are also land gastropods. They have *tiny internal shells* and relish tender garden produce.

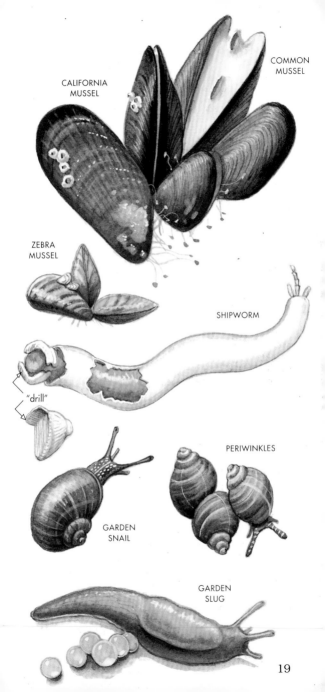

CALIFORNIA MUSSEL

COMMON MUSSEL

ZEBRA MUSSEL

SHIPWORM

"drill"

GARDEN SNAIL

PERIWINKLES

GARDEN SLUG

19

ANNELIDS

These moisture-loving worms have segmented bodies. They include earthworms, leeches, and marine worms.

EARTHWORM   7 in.

These familiar worms continuously till and aerate the surface of the Earth, eating tiny organisms from the soil and emerging at night to browse on vegetable matter. The "castings" they excrete enrich the soil. Each segment of an earthworm's body contains tiny bristles, or *setae*, that help it move.

TUBIFEX WORM   To 1 in.

Often used as live fish food by those who keep aquariums, *red* tubifex worms thrive in polluted ponds and other fresh water.

POND LEECH   To 5 in.

These freshwater members of the "two-suckered worm" family have *suckers at both ends*. They are blood-drinking predators that can be found attached to turtles—or to us. Their bite is not painful, however, because the saliva of leeches contains a natural anesthetic. They can be encouraged to let go by the application of heat, salt, or alcohol. Leech saliva also has a substance that keeps blood from clotting, which makes leeches interesting to medical researchers.

TARDIGRADE $1/40$ in.

Mosses, gutters, and plant debris—any place that is wet and dry by turns—contain large populations of these charming creatures. Also called "water bears," they have *eight legs* and a lumbering gait, and they feed themselves by sucking the juices of plant cells. Though tardigrades look cute, they are extremely hardy little animals. They can withstand large doses of radiation or temperatures approaching absolute zero, and they can remain dormant for over 100 years awaiting favorable conditions. When times get really hard, a water bear can produce a capsule called a tun on which it rides the breezes to a new home.

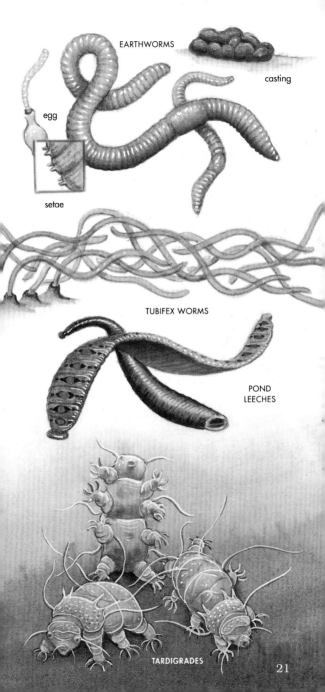

EARTHWORMS

casting

egg

setae

TUBIFEX WORMS

POND
LEECHES

TARDIGRADES

21

ARTHROPODS

Nearly 80 percent of all animal species are arthropods, and many have easily made themselves at home in human habitats. Scorpions and spiders, insects and centipedes, crabs and lobsters are all arthropods.

SCULPTURED SCORPION   2 3/4 in.

This Arizona native is our most poisonous scorpion. There are 70 other scorpion species in North America, most of them in the South and Southwest. All have *stingers* at the ends of their tails. Scorpions hide from the sun under trash or in crawl spaces, shoes, or clothing. If a scorpion crawls on you, brush it off, don't swat it.

BROWN RECLUSE SPIDER   3/8 in.

All spiders have fangs and venom for killing other arthropods. Most spiders' venom is too mild to bother humans, but a few have a more powerful bite. The Brown Recluse Spider is a small, *venomous, light brown* spider that may hide in buildings, woodpiles, or folded cloth, usually in the South. It has a small *violin-shaped mark* on its head.

BLACK WIDOW SPIDER   3/8 in.

Beautiful and notorious, the Black Widow is occasionally found in houses in warmer parts of North America. It is *black* with a *red mark* on its abdomen. The bite of the Black Widow can be extremely painful.

AMERICAN HOUSE SPIDER   1/4 in.

Harmless and helpful, the marble-patterned American House Spider is our most common indoor spider. Its sticky cobwebs trap insects—and dust. Like its cousin the Black Widow, the female produces a pear-shaped egg case, which she guards faithfully.

GARDEN SPIDER   3/4 in.

Where there is a garden with insects to eat, you will find this handsome orb-weaving spider with *white spots*. Like many other orb-weavers, every day it eats its old web and weaves a new one.

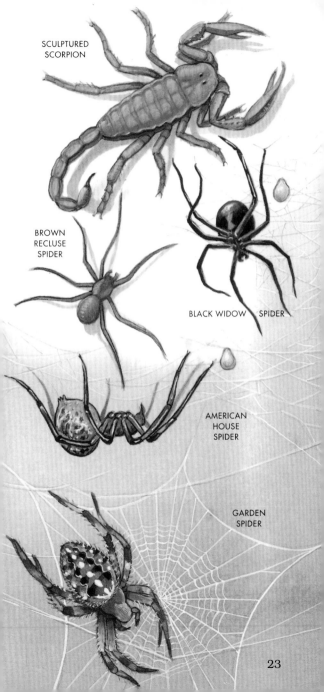

SCULPTURED
SCORPION

BROWN
RECLUSE
SPIDER

BLACK WIDOW   SPIDER

AMERICAN
HOUSE
SPIDER

GARDEN
SPIDER

23

LITTLE WHITISH HOUSE SPIDER   3/8 in.

The Little Whitish House Spider, with its *long front legs,* has increased its range in the north by living in our warm buildings. It is reclusive and may be responsible for unexplained bites that cause a slight fever and are slow to heal.

DARING JUMPING SPIDER   3/8 in.

Most jumping spiders live outdoors, but the Daring Jumping Spider, black and white with metallic green jaws, lives in buildings. Like all jumping spiders, this stocky little spider has *short legs* and sees well with its *large eyes.* Jumping spiders do not spin webs but let out a silk drag line when they fall or leap upon their prey. Many jumping spiders are very colorful. Males wave their front pair of legs when courting females.

HUNTSMAN SPIDER   1 in.

The large Huntsman Spider lives in Florida. It is beneficial to households that are broad-minded enough to accept its presence, because it hunts cockroaches and other domestic insects.

DADDY LONGLEGS   1/4 in.

This arthropod is not a true spider but a member of the "harvestman" group. It eats tiny pieces of its prey instead of drinking their juices as true spiders do. Harvestmen also lack the slender "waist" of true spiders. Their delicate legs break off easily and do not grow back. Most Daddy Longlegs live outside, but some kinds prefer the darkness of our cellars.

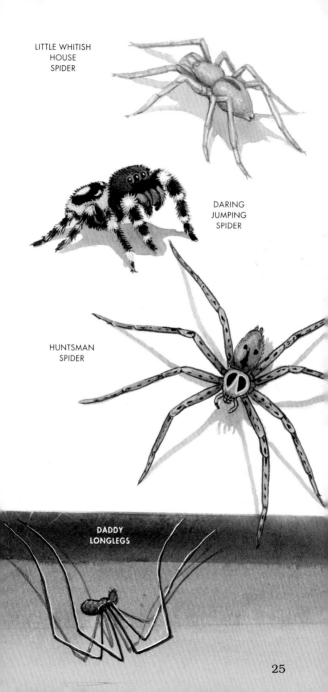

LITTLE WHITISH
HOUSE
SPIDER

DARING
JUMPING
SPIDER

HUNTSMAN
SPIDER

DADDY
LONGLEGS

25

Mites and Ticks

Adult mites and ticks have four pairs of walking legs and feed on fluids, like spiders. These tiny arthropods prey on plants and animals and cause many health problems.

DUST MITE

These minute creatures—so small that 30 could sit together on the head of a pin—exist in unimaginable numbers in dust, carpets, and bedding, where they consume microscopic bits of our skin and spilled food. Many people are allergic to Dust Mites.

PREDATORY DUST MITE

Fortunately, Predatory Dust Mites eat house dust mites. Unfortunately, our vacuum cleaners may suck up more predators than prey, making the situation worse.

FOLLICLE MITES

These mites live harmlessly in the follicles of our eyelashes and body hairs, feeding on body oils and hanging on with eight stubby legs.

GRAIN MITE

The equally tiny Grain Mite feeds on grain products everywhere. It needs the help of hungry fungi to break down its food.

CHIGGER $^1/_{10}$ in.

Small, *bright red*, and beneficial, adult Chigger mites feed on insects. The almost invisible larvae, however, feed on us. They bite in warm, moist places where clothing is tight.

HUMAN ITCH MITE $^1/_{10}$ in.

The Human Itch Mite burrows in creases in our skin, causing a fierce itch called scabies. A related mite causes mange in dogs.

BROWN DOG TICK $^1/_5$ in.

An engorged Brown Dog Tick (size given is *before* dinner) is a startling sight. Ticks wait on leaves and grass for a blood meal to stroll by. Ticks can transmit serious diseases.

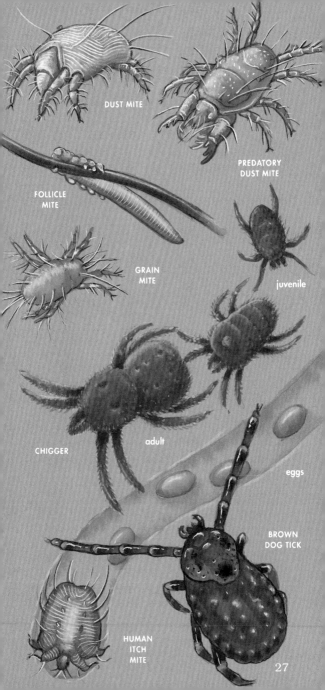

DUST MITE

PREDATORY DUST MITE

FOLLICLE MITE

GRAIN MITE

juvenile

CHIGGER

adult

eggs

BROWN DOG TICK

HUMAN ITCH MITE

27

Millipedes and Centipedes

These arthropods have long bodies with many segments and numerous legs. The nonpoisonous millipedes have two short pairs of legs per body segment. The predatory centipedes have one long pair of legs per body segment.

HOUSE MILLIPEDE $1^1/_2$ in.

Most millipedes are vegetarians and scavengers, living outdoors under moist debris. The house millipede sometimes lives with us. Millipedes protect themselves by releasing a foul odor. Some can also roll into a tight ball.

HOUSE CENTIPEDE $1^1/_4$ in.

The nippy house centipede helps us by feeding on insects in our homes. During its nightly patrols it sometimes gets caught in the bathtub.

Insects

The insects are by far the largest and most successful class of arthropods. They have a three-part body and three pairs of walking legs, and most have wings.

SILVERFISH $^1/_2$ in.

Triple-tailed, pearly colored, tapered and quick, silverfish live in our homes, libraries, and other buildings, eating all sorts of starchy material, from glues used in bookbinding to spilled food. They prefer cool, damp places like the bathroom.

FIREBRAT $^1/_2$ in.

More heat-loving than the silverfish is its cousin, the Firebrat, which is found near stoves, furnaces, and fireplaces. It has *darker markings* than a silverfish but also eats starch.

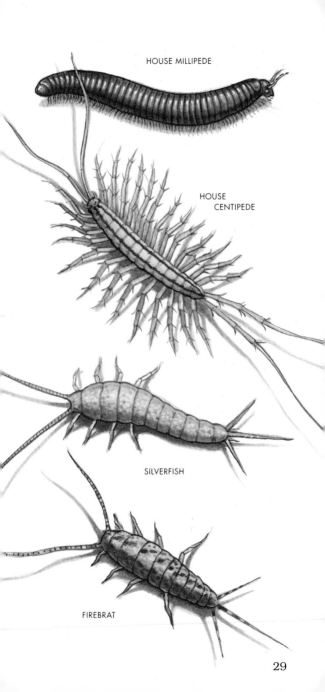

HOUSE MILLIPEDE

HOUSE
CENTIPEDE

SILVERFISH

FIREBRAT

29

MAYFLY $^7/_8$ in.

Mayflies are notable for their *long, double* or
triple tails; their transparent, fragile wings;
their massive hatches; and their very brief
adult lives. The larvae, also called nymphs,
live in fresh water. The adults emerge all
together in awesome, harmless, gossamer
swarms, sometimes covering streets and
windows with their delicate bodies. The
adults live only a day or two and never even
eat. They have only one goal: to mate and
lay eggs.

GREEN DARNER 3 in.

To some people, the sharp-looking bodies,
large eyes, and sometimes rattling flight of
dragonflies and damselflies are frightening.
But although they are hunters, they are not
looking for us. Zooming over fresh water,
they seek mosquitoes and other insect prey.
A very few of the larger species can bite if
provoked. The Green Darner dragonfly is
common across North America. Like most
dragonflies, it holds its wings out from its
sides when at rest.

STOCKY DAMSELFLY $1^1/_2$ in.

The Stocky Damselfly is common across all
but the most southern parts of North America. Like most damselflies, it holds its wings
folded over its back while resting. Its
nymphs (larvae), like the much stockier,
fiercer dragonfly nymphs, hunt for food
underwater.

MAYFLY

Mayfly nymph

GREEN
DARNER

STOCKY DAMSELFLY

Damselfly
nymph

Dragonfly
nymph

31

*Cockroaches have lived on Earth virtually un-
changed for more than 350 million years. They
are normally tropical insects, but we have
obliged a few species by providing warm,
moist, food-filled homes for them to infest.
Tough, fast, prolific, and able to learn to avoid
poisons, cockroaches often survive our efforts to
eradicate them. Although winged, these oval,
flat-bodied insects rarely fly. They are not directly implicated in the spread of diseases, but
their feces and particles from their bodies cause
troublesome allergies in many people.*

AMERICAN COCKROACH   2 in.
The American Cockroach, with *antennae
longer than its body*, is one of our largest
roaches. This species is often called the
sewer roach.

ORIENTAL COCKROACH   1³/₈ in.

This *blackish* Asian cockroach has a body
that looks slightly more *beetlelike* than
other roaches. It is usually found in base-
ments where it has access to water.

BROWN-BANDED COCKROACH   ⁵/₈ in.
The Brown-Banded Cockroach lives all over
the house in odd places, like inside the tele-
vision. It tends to fly more readily than
other species. Its sticky egg cases, which
like all cockroach egg cases look like little
purses, are sometimes found on walls and
other vertical surfaces.

GERMAN COCKROACH   ⁵/₈ in.
The very common *brown* German Cockroach seeks water and warmth, so it is often
seen in the bathroom. It too likes televisions
and other warm appliances, which not only
provide warm hiding places but supply
snacks in the form of various pastes and
coatings used during manufacturing.

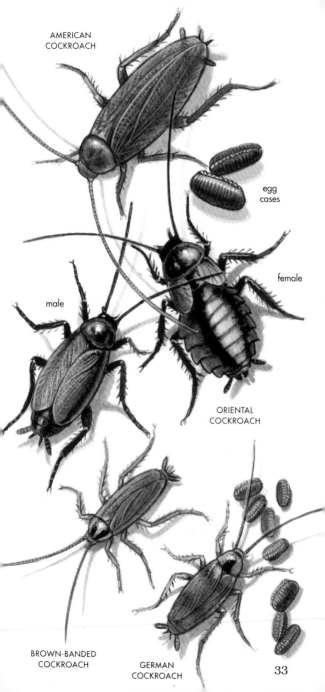

AMERICAN
COCKROACH

egg
cases

male

female

ORIENTAL
COCKROACH

BROWN-BANDED
COCKROACH

GERMAN
COCKROACH

33

PRAYING MANTIS   3 in.

Formidable and impressive, the Praying Mantis and other members of the mantid group are easy to recognize. Both adult and young mantids are fierce insect predators. All mantids have *strong, spiny front legs* and long bodies. The Praying Mantis has a dark ringed *yellow spot* in its armpit.

SUBTERRANEAN TERMITE   3/8 in.

Termites are *wingless*, except for mating adults, and have pale, soft bodies shaped like bowling pins. Only breeding adults briefly have wings. Most termites are tropical, but a few North American species, like the Subterranean Termite, can invade the wood of our houses, seriously weakening the timbers.

RING-LEGGED EARWIG   1 in.

The tarry-smelling earwigs look menacing, with a *pair of large cerci*, or pincers, at the ends of their bodies. Earwigs can pinch, but they are nonvenomous and do not bite. Nocturnal and omnivorous, earwigs do not enter ears, but they do like crevices under litter or in buildings. The Ring-legged Earwig can be found in most parts of North America.

RED-LEGGED GRASSHOPPER   1 in.

The "singing" of grasshoppers heralds the heat of summer. For gardeners and farmers, these jumping insects can be destructive pests, because they eat crops and can occur in huge numbers. This *black-spined* grasshopper can be found in vacant lots in cities.

HOUSE CRICKET   3/4 in.

The *brownish* House Cricket is found indoors all over the world. It creates its persistent chirp by scraping the sharp edge of one front wing against filelike ridges on the other front wing. It hides during the day and hunts for food scraps at night.

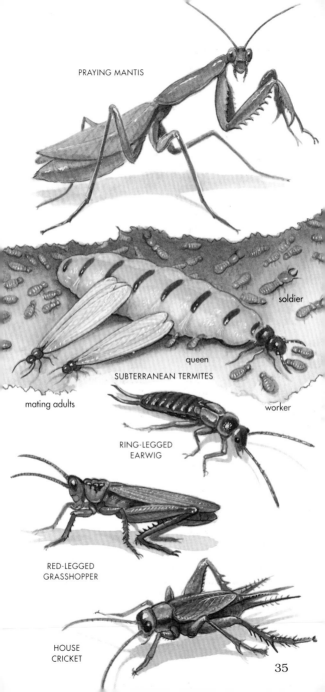

PRAYING MANTIS

SUBTERRANEAN TERMITES

soldier

queen

mating adults

worker

RING-LEGGED
EARWIG

RED-LEGGED
GRASSHOPPER

HOUSE
CRICKET

35

BOOKLOUSE   1/8 in.

Booklice are not true lice, but pale, *wingless* insects that usually feed on organic matter such as book paste. They look a little like tiny termites with protruding eyes. The Booklouse can infest houses, eating wallpaper and book pastes.

HUMAN BODY LOUSE   1/10 in.

True sucking lice live on their hosts' blood. Some can transmit diseases like typhus. The surprisingly fast, pinkish Body Louse, or "cootie," parasitizes humans. A close relative is the Head Louse, which most commonly infests children.

CRAB LOUSE   1/12 in.

The Crab or Pubic Louse clings to the hairs of our private parts with great tenacity. It is the only louse besides Head and Body Lice that parasitizes humans.

BEDBUG   1/4 in.

Many people incorrectly call all insects "bugs." True Bugs have sucking mouthparts and partly transparent forewings that they fold over their backs. They include water bugs, stink bugs, and many others, as well as the Bedbug. This *flat, reddish* insect is less common than it used to be in North America, but it still troubles 80 percent of the world's people. Its bites leave tiny, itchy, red bumps on the skin.

CICADA   1 3/8 in.

Husky cicada larvae live underground for as long as 4 to 20 years, then emerge and climb into trees to molt, mate, and finish their brief adult lives. Male cicadas "singing" to attract mates are the loudest of insects.

SPITTLEBUG   1/3 in.

A frothy mass of bubbles on green stems protects the tiny Spittlebug nymphs while they feed on plant juices. The hopping, mottled brown adults are called Froghoppers.

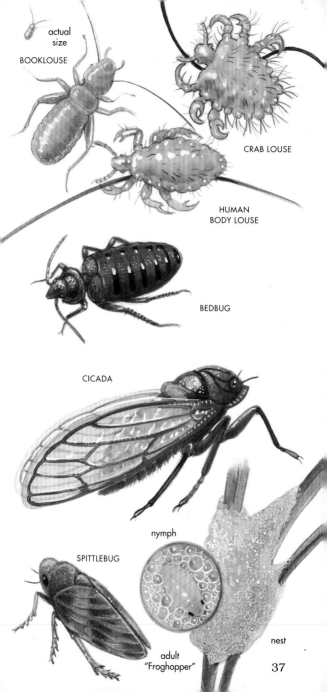

actual size

BOOKLOUSE

CRAB LOUSE

HUMAN
BODY LOUSE

BEDBUG

CICADA

SPITTLEBUG

nymph

adult
"Froghopper"

nest

37

WHITEFLY $^1/_{15}$ in.

Common pests of tender household plants, whiteflies look like tiny, sometimes flying, *white specks*. Nymphs and adults suck juices from the undersides of leaves, secreting a sticky honeydew that, in turn, attracts a fungus that blocks the "breathing pores" of the leaves.

APHID $^1/_{16}$ in.

Aphids also feed on plant sap and can be extremely damaging to farm crops as well as plants in our houses and gardens. They have tiny, pear-shaped bodies and appear in many colors.

COTTONY CUSHION SCALE $^1/_3$ in.

Destructive, juice-sucking Cottony Cushion Scale insects stay in one spot on their host plants, covered in whitish wax. Females protect their eggs with soft, white "pillows" of fibers.

MEALYBUG $^1/_8$ in.

Disguised with a heavy layer of *grayish wax*, slow-moving mealybugs also have several long *taillike filaments*. They encrust and damage many plants.

ANTLION $1^3/_4$ in.

The adult Antlion has clublike antennae and large, transparent wings that it holds over its long body. The more familiar larva, also known as the "doodlebug," digs funnel-shaped pits in open, sandy areas, then lies buried at the bottom of the pit with only its mouthparts showing. There it waits for ants to stray over the edge and down the slippery sides to their doom in the jaws of the voracious doodlebug.

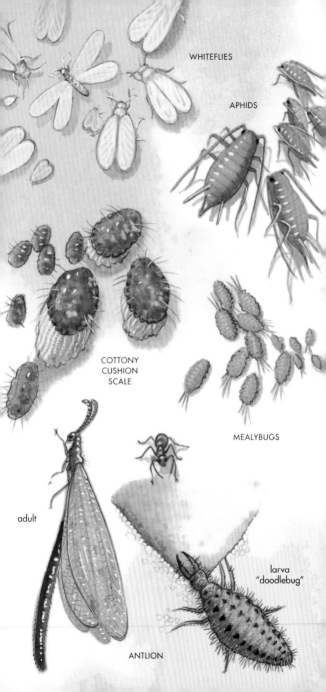

WHITEFLIES

APHIDS

COTTONY
CUSHION
SCALE

MEALYBUGS

adult

larva
"doodlebug"

ANTLION

COMMON BLACK GROUND BEETLE $^5/_8$ in.
Found throughout North America, this and other ground beetles often hide under logs or old boards. Although beetles tend to crawl rather than fly, they do have wings; their forewings form protective covers for their seldom-used hind wings.

JAPANESE BEETLE $^1/_2$ in.
This iridescent insect is a member of the handsome, sturdy, and often colorful scarab beetle family. It is a real pest in the garden.

JUNE BEETLE $1^3/_8$ in.
Drawn to our lights and bumbling against our screens is another scarab, the night-flying, large, brown June Beetle. The whitish larvae, or grubs, of Japanese and June beetles live in the ground.

FIREFLIES $^1/_2$ in.
East of the Rocky Mountains are several kinds of fireflies with luminous, flashing glands at the tips of their bodies that attract females. Tall grass and a lack of pesticides encourage these magical beetles.

TWO-SPOTTED LADY BEETLE $^1/_4$ in.
The round little "lady bug" feeds on aphids in the garden, and often finds its way into our houses in fall. It is *orange* with *black and white* markings.

CONFUSED FLOUR BEETLE $^1/_8$ in.
The larvae of this red-brown beetle are wiry and yellowish white. The Confused Flour Beetle can be a serious pest.

SAWTOOTHED GRAIN BEETLE $^1/_8$ in.
This *flat*-bodied insect slips into food packages to eat and lay its eggs.

DERMESTID BEETLES $^1/_8$ in.
The large, bristly larvae of these beetles scavenge in pantries and also eat wool, feathers, and fabrics.

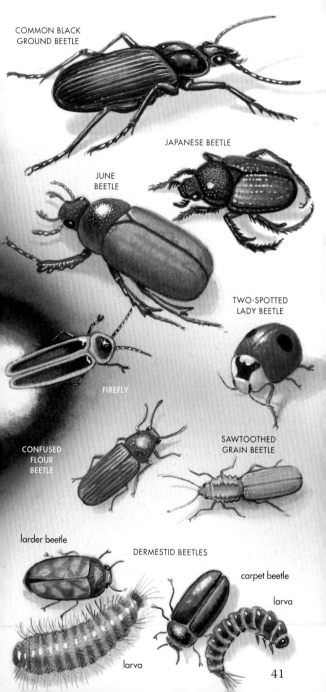

COMMON BLACK
GROUND BEETLE

JAPANESE BEETLE

JUNE
BEETLE

TWO-SPOTTED
LADY BEETLE

FIREFLY

CONFUSED
FLOUR
BEETLE

SAWTOOTHED
GRAIN BEETLE

larder beetle

DERMESTID BEETLES

carpet beetle

larva

larva

larva

41

FLEA $^1/_{16}$ in.

More people have been killed by flea-trans-
mitted disease, including bubonic plague
and typhoid, than have died in all the wars
ever fought. Like all fleas, the Dog Flea and
the Cat Flea are wingless, hard, *flat,* and
jumpy. They drink blood, including our
own, and scatter their eggs and larvae in
and under upholstery and rugs, even in the
tidiest houses. Flea bites itch much longer
than mosquito bites.

MOSQUITO $^1/_4$ in.

Sweat, exhaled carbon dioxide, and warm,
moist air attract mosquitoes. Only females
bite, to get the blood meal they need before
they lay their eggs in stagnant water. Males
sip plant juices. Diseases like encephalitis
and Lyme disease are occasionally trans-
mitted by the female's saliva, which can also
carry heartworm nematodes to dogs.

FRUIT FLY $^1/_{16}$ in.

Tiny fruit flies seem to appear out of
nowhere anywhere there is overripe fruit.
They come to feed on the yeasts that, in
turn, are feeding on the softening fruit.

HOUSE FLY $^1/_4$ in.

The familiar House Fly has distinct *stripes*
on its back and *pale sides* to its abdomen.
Flies eat and lay their eggs in manure, rotting vegetable matter, and dead flesh, and
so transmit many diseases including cholera and dysentery. Fly larvae (maggots),
however, are valuable recyclers of decaying
organic matter.

FLESH FLY
BLUEBOTTLE FLY $^1/_2$ in.

The Bluebottle has a *metallic blue abdomen*;
the Flesh Fly is striped. Both flies sometimes enter houses seeking meat in which
to lay their eggs.

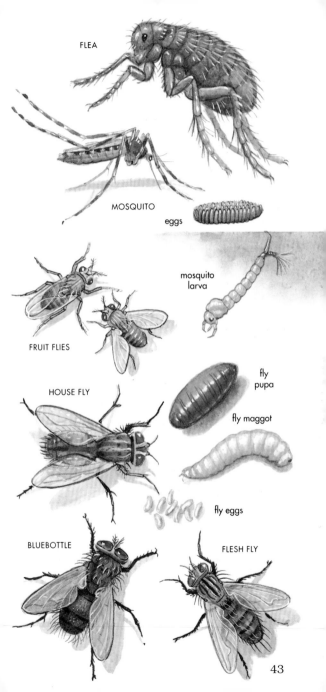

FLEA

MOSQUITO

eggs

FRUIT FLIES

mosquito larva

HOUSE FLY

fly pupa

fly maggot

fly eggs

BLUEBOTTLE

FLESH FLY

43

*Butterflies and moths are among the most
noticeable of the insects that live with us. Butterflies normally fly during the day and have
clubbed antennae; moths usually fly at night
and often have feathery antennae. In warm
weather you can always find moths near city
lights at night.*

CLOTHES MOTH $^1/_4$ in.
Like most moths, a small, pale, light-shy
clothes moth rests with its *fringed wings*
folded down against its body. The small,
whitish caterpillars eat not only wool but
also furs and even the felt inside pianos.

CABBAGE BUTTERFLY 2 in.
The *white* Cabbage Butterfly with its *black-
spotted* wings lays its eggs on plants in the
cabbage family. The eggs hatch into green
caterpillars with a lengthwise yellow stripe
that eat big holes in the leaves.

MOURNING CLOAK $3^3/_8$ in.
The handsome Mourning Cloak butterfly is
an iridescent *purple-black*, with golden
edges to its wings. It is common in developed areas, and lays its eggs on elm, willow,
and poplar trees. These are the favorite food
plants of its velvety black caterpillar, which
has branched spines and 8 red spots.

MONARCH 4 in.
The Monarch butterfly, *orange* with a *triple
row of white spots* in the black margins of
its wings, is famous for its 2,000-mile winter migrations to Mexico. Its caterpillars
feed exclusively on milkweed plants, which
produce toxins that taste terrible, good pro-
tection from predators for both the caterpillar and the adult.

VICEROY 3 in.
The Viceroy butterfly mimics the Monarch's
colors, thereby benefiting from the Monarch's bad reputation, but some Viceroys
are bad-tasting too. The Viceroy has slim,
curved *black bands* across its hindwings.

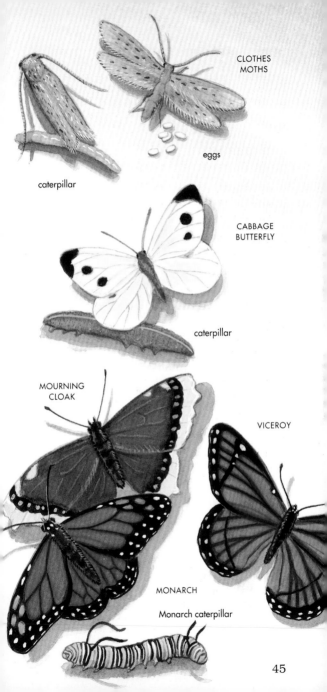

CLOTHES
MOTHS

eggs

caterpillar

CABBAGE
BUTTERFLY

caterpillar

MOURNING
CLOAK

VICEROY

MONARCH

Monarch caterpillar

45

INDIAN MEAL MOTH   5/8 in.
MEAL MOTH   1 in.

The Indian Meal Moth, with its half-gray, half-dusky forewings, and its cousin, the Meal Moth, with its brown- and beige-banded forewings, slip into household grain products to lay their eggs. Their small caterpillars feed there, betraying their presence by little webs that bind grains or other food together.

TENT CATERPILLAR MOTH   1 1/2 in.

The bristly, blue-striped caterpillars of the Tent Caterpillar Moth build their silken homes in tree branches. The adult moth has *two light stripes* on each forewing that form a double V when the moth rests with its wings folded back over its body.

LUNA MOTH   4 1/2 in.

We are lucky in North America to have several members of the big, dramatically marked giant silkworm moth family present even in developed areas. These moths are only distantly related to the true silk-producing moths of Asia. The *lime green* Luna Moth has long, elegant *tails* on its hindwings. The presence of Luna Moths is a sign of a healthy habitat, for they are very sensitive to pesticides.

CECROPIA MOTH   5 7/8 in.

Surprisingly, the huge Cecropia Moth is often found in populated areas, attracted to our lights and to many species of urban street trees. It has *scalloped, whitish borders* at the outside edges of its forewings. Its body is *red* with a *white collar* and *white stripes* around its abdomen.

CYNTHIA MOTH   5 3/8 in.

The Cynthia Moth has followed its Asian host tree, the tree-of-heaven or ailanthus, wherever it has spread in urban areas across North America. The Cynthia Moth is *olive* green or brownish with bold white markings.

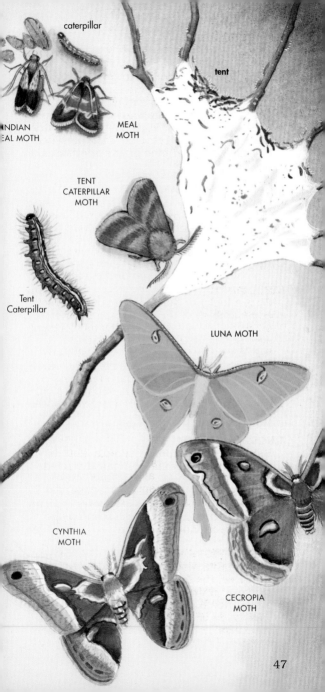

caterpillar

INDIAN
MEAL MOTH

MEAL
MOTH

tent

TENT
CATERPILLAR
MOTH

Tent
Caterpillar

LUNA MOTH

CYNTHIA
MOTH

CECROPIA
MOTH

47

Some insects of the ant, wasp, and bee group are highly social, with many generations and different kinds of workers living together for their mutual benefit.

VELVET ANT   1 in.
This insect is not really an ant but a wasp—a *hairy, red, wingless* female looking for insect pupae to parasitize with her eggs. Also called "cow killers," velvet ants pack a painful wallop of a sting. Male velvet ants have wings and no stingers.

PHARAOH ANT $1/16$ in.
Warmth-loving and nonbiting, *tiny, brown* pharaoh ants are found almost anywhere people give them a home. Large colonies simply divide and, linked by odor trails, establish sprawling communities of millions of workers and many queens.

LITTLE BLACK ANT $1/16$ in.
The Little Black Ant nests outside our homes, below ground, with the colony entrances marked by small craters. In the South it occasionally moves indoors. It likes sweet foods.

CARPENTER ANT $1/2$ in.
East of the Rockies, *large, black* carpenter ants (bigger and light brown in Texas) nest in holes gnawed in dead wood, including our homes, where they can do a lot of damage. In large numbers, they smell of formic acid, which they use as a weapon and a defense. A nip with a spritz of formic acid in it makes a painful, hard swelling.

PAVEMENT ANT $1/8$ in.
Brown, hairy pavement ants usually nest under stones, asphalt, and concrete. They often invade houses, especially in summer.

ODOROUS HOUSE ANT $1/8$ in.
The brown to black Odorous House Ant travels in lines. When other food sources fail, it enters our homes. If crushed, it gives off a *coconut odor*.

48

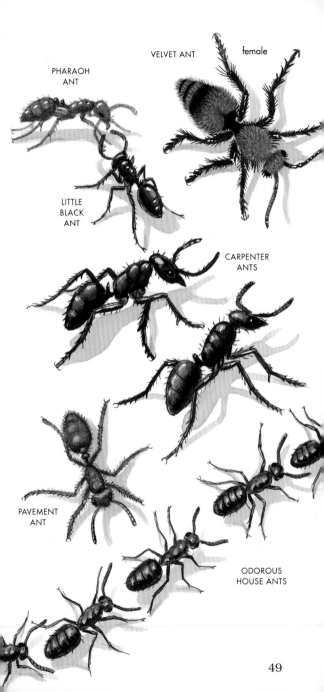

VELVET ANT

female

PHARAOH ANT

LITTLE BLACK ANT

CARPENTER ANTS

PAVEMENT ANT

ODOROUS HOUSE ANTS

49

PAPER WASP   1 in.

Paper wasps can sting, but they are less irritable than hornets and Yellowjackets. They are slender-bodied and *brown* with *narrow yellow and black rings*. Adults drink nectar and juice from rotting fruit. Females tend eggs and larvae in delicate, papery nests made of wood pulp and gluey saliva.

BALD-FACED HORNET $^3/_4$ in.

This *black and white* wasp is a short-fused animal that can sting repeatedly. Bald-Faced Hornets build large, hanging, enclosed nests out of chewed wood, grass, and cardboard. They forage for nectar and softening fruit at edges of woods.

YELLOWJACKET $^5/_8$ in.

Yellowjackets are the wasps most likely to attend your picnics. Their nest is virtually the same as the Bald-Faced Hornet's but is normally built underground. They are protective of their territory; accidentally stepping on a nest is a memorable experience.

HONEY BEE $^5/_8$ in.

The *striped* golden brown, black, and orange-yellow Honey Bee you see is probably a nonbreeding female worker. She is able to communicate the location of nectar sources to her hivemates by a complex dance. Honey Bees are relatively peaceable and can sting only once.

BUMBLEBEE $^7/_8$ in.

Bumblebees are armed with stingers, but they are not aggressive. True to their name, they are the larger, clumsier, and noisier bees we see tumbling about in flowers. This *stout* and *hairy*, yellow and black bee nests in small underground colonies.

CARPENTER BEE   1 in.

In spring months, you may see a *very large*, *black* bee studying your wooden house. It is a female carpenter bee, looking for a promising place to chew a foot-long tunnel in which to lay her eggs.

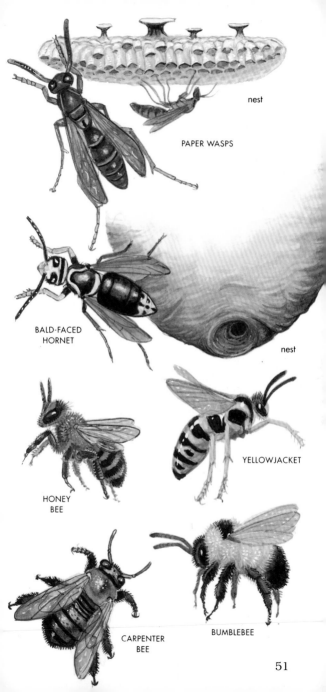

nest

PAPER WASPS

BALD-FACED
HORNET

nest

HONEY
BEE

YELLOWJACKET

CARPENTER
BEE

BUMBLEBEE

51

Crustaceans

Most of these arthropods have gills and live in the water. There are a very few land-living crustaceans.

ROCK BARNACLES   1 in.
These crustaceans are disguised as mollusks. Attached firmly to coastal rocks and pilings, they lie in overlapping shells, rhythmically moving their legs through the water like sieves to catch tiny prey.

NORTHERN LOBSTER   16 in.
Native to the coastal waters of the Northeast, the Northern Lobster sulking in the supermarket is encased within a jointed exoskeleton. It has a large, blunt-tipped claw for crushing shellfish and a smaller, sharp-tipped claw for tearing food to pieces. Live lobsters are *blue-green;* they turn red when cooked.

DUNGENESS CRAB   9 in.
BLUE CRAB   9 in.
ROCK CRAB   5 in.
The *dusky* Dungeness Crab from the West Coast, the *speckled, reddish* Rock Crab from the East Coast, and the *blue-legged* Blue Crab from brackish waters of the East Coast end up in supermarkets too.

GRIBBLES $1/32$ in.
Wherever there is unprotected wood in cool salt water, there are countless numbers of tiny crustaceans called gribbles. They graze on the wood and organisms that live on and in it. Gribbles and shipworms (see p. 18) together do great damage to dock pilings; gribbles chew from the outside while shipworms work from within.

PILLBUG $5/8$ in.
SOWBUG $5/8$ in.
In the cellars of old buildings and under logs and yard debris, you can almost always find these land-living crustaceans. The shiny pillbugs can *roll into a ball* in self-defense. The *flatter,* paler sowbugs cannot.

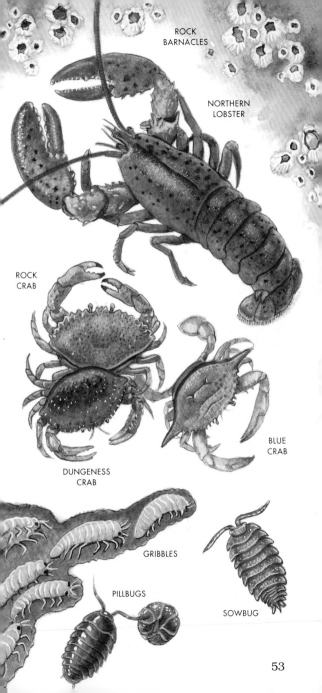

ROCK BARNACLES

NORTHERN LOBSTER

ROCK CRAB

DUNGENESS CRAB

BLUE CRAB

GRIBBLES

PILLBUGS

SOWBUG

53

ECHINODERMS

These ocean-dwelling animals have radial symmetry, meaning they have several (usually five) similar parts radiating from a central hub. They include the spiny-skinned sea stars, sea urchins, and their kin. Echinoderms are thought by some to be organisms that gave up a livelier life early in their evolutionary history in favor of a more sedentary one. Their larvae are similar in some ways to the embryos of animals with backbones. This means that sea stars may be a little more closely related to humans than they are to animals without backbones, such as arthropods. With this provocative thought in mind, we can admire them also for their slow-moving mastery of the element in which they live: water. They move by controlling the water pressure inside their bodies—without any brains at all.

NORTHERN SEA STAR   5 in.

These animals are commonly known as starfish, but they are not really fish at all. On the East Coast, *reddish* Northern Sea Stars can be seen by the hundreds hunting for mussels, their favorite food, on rocks and pilings and in tide pools. The *pale spot*, called a madreporite, controls the flow of water that lets the sea star move its feet. If a predator takes an arm or two, the limbs will grow back.

LEATHER STAR   5 in.

The garlicky smelling, *leathery* feeling Leather Star lives on the West Coast. Like all echinoderms, the Leather Star uses water pressure to coordinate the actions of its hundreds of tube feet. The Leather Star moves slowly on sea walls, rocky shores, and pilings in search of anemones and sea cucumbers to eat.

NORTHERN
SEA STAR

LEATHER STAR

55

Fishes

There are more fish species than there are species of any other backboned animal. A few can tolerate urban conditions and are found swimming through our cities' rivers or in our harbors.

GOLDFISH   To 16 in.

The wild Goldfish is a dull coppery brown color, quite unlike the gaudy pet-store varieties. Goldfish are messy bottom-feeders, uprooting aquatic plants and smothering other life forms with the mud they stir up. They are relatively tolerant of polluted waters. Goldfish were first domesticated in China 10 centuries ago; there are now over 125 breeds of pet goldfish, including some bizarre forms like the "bubble eye" shown at right. Studies have shown that goldfish can solve intelligence tests as well as rats can. Pet goldfish released in the wild will reproduce rapidly, but the colorful young are quickly eaten. Brownish offspring have a better chance at survival, so the population will quickly revert to the dull colors of the wild fish. Pets should never be released in the wild; they will either suffer or become disruptive pests.

COMMON CARP   To 3 ft.

Carp are brassy green fish with large scales and a long, reddish dorsal fin and tail. Introduced widely in North America in the 19th century, the Carp is now considered a "trash fish" here, one that roots in mud and stirs up streams and ponds so that native species cannot survive. Breeding pairs are the usual cause of languid thrashings at the surface of weedy ponds in summer. In Japan and England, Carp are highly valued for their long lives, uncanny wariness, and tough fighting spirit. Japanese Koi (shown at right) are carp that have been selectively bred for color and form. In Japan, a single prize-winning Koi may sell for $40,000.

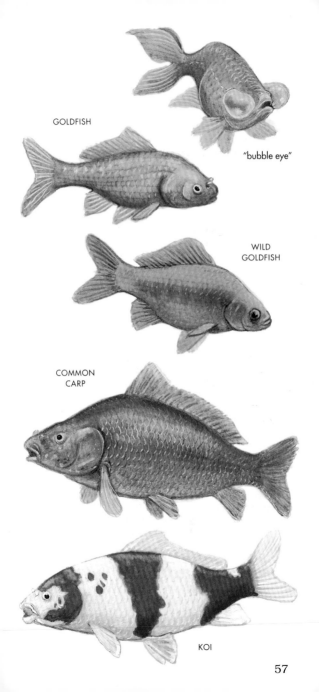

GOLDFISH

"bubble eye"

WILD
GOLDFISH

COMMON
CARP

KOI

57

BLACK BULLHEAD **To 16 in.**

Like all catfishes, this one has "whiskers" (harmless, taste bud-covered barbels) as well as small eyes and acute hearing, all design features that help them forage for food at night or in murky water. Many catfishes have mildly poisonous spines in their fins. The Black Bullhead is commonly— almost willingly—caught in muddy ponds and river bottoms in the central United States. This *dusky-backed, yellowish-bellied* fish with dark barbels can tolerate the polluted conditions in urban areas.

YELLOW BULLHEAD **To 17 in.**

This particularly slippery catfish has an olive brown back, yellow belly, and *white* barbels. It is fond of quiet waters with much vegetation, but occasionally enters urban waterways in the Midwest and east to the coast.

CHANNEL CATFISH **To 4 ft.**

The fork-tailed Channel Catfish has a mild taste that has made it a favorite of fish farmers and catfish lovers in the South. The "speckled cat" is spattered with *dark spots* until is quite large. It is occasionally fished from river waterfronts of less polluted developed areas.

TADPOLE MADTOM **To 5 in.**

Small and dark, with an oversized tail, this chubby little catfish looks like a tadpole as its dashes from one hiding place to another. It is most likely to be found under logs in mud and vegetation when it enters urban waterways in the Midwest and along the East Coast.

WALKING CATFISH **To 14 in.**

If you live in Florida, you may be startled to find this slender, sometimes albino catfish wriggling through the grass on a rainy night as it travels from one fresh waterway to another. Native to Asia, released or escaped Walking Catfishes sometimes become established in the wild.

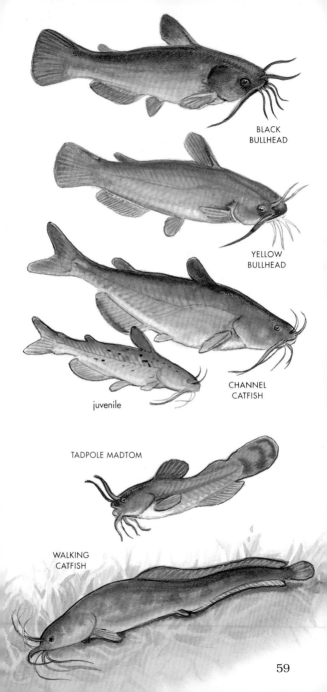

BLACK
BULLHEAD

YELLOW
BULLHEAD

CHANNEL
CATFISH

juvenile

TADPOLE MADTOM

WALKING
CATFISH

59

PUMPKINSEED   To 7 in.

This is a member of the large sunfish family.
Male "sunnies" are among the most brightly
colored fish in North America and have long
provided fishing excitement for children in
the East and upper Midwest. This small-
mouthed, *blue-green* sunfish has *red, white,
and black "ears,"* a *golden yellow belly,* and
orange flecks on its body. It likes sluggish
water with lots of vegetation and can be
found in many urban ponds.

GREEN SUNFISH   To 7 in.

Very common, this *large-mouthed, large-
headed, olive green* sunfish tolerates a wide
range of conditions throughout the greater
Mississippi drainage area. It is an even
more accomplished bait stealer than most
members of the sunfish tribe.

BLUEGILL   To 9 in.

The dark-banded Bluegill is a popular sport
fish and is widely stocked outside its native
regions. Like most sunfish, the male Bluegill clears a pale patch in the weeds of a
pond bottom in which the female lays her
eggs. He then hovers over it, aggressively
guarding eggs and young.

REDEAR SUNFISH   To 9 in.

Like the Pumpkinseed, the Redear also has
a bright *red spot* on its black ear flap. It is
found throughout the Deep South, west into
Texas and up into the southern Midwest,
except where pollution has killed the snails
on which it feeds.

LARGEMOUTH BASS   To 3 ft.

Feeding on other fishes, frogs, and an occasional duckling, this large member of the
sunfish family can tolerate warm and weedy
water. It is much sought after as a game fish
and has been widely introduced. There are a
number of regional varieties, but all have a
dusky stripe or a *row of spots* from head to
tail. It indeed has a *large mouth.*

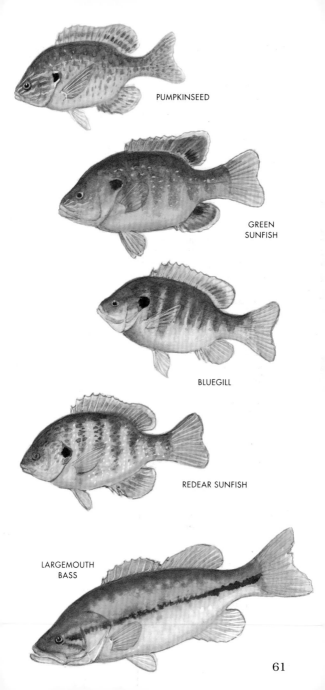

PUMPKINSEED

GREEN
SUNFISH

BLUEGILL

REDEAR SUNFISH

LARGEMOUTH
BASS

61

SHEEPSHEAD   To 3 ft.
The *banded* Sheepshead is a member of the
stout-spined, large-headed Atlantic porgy
family, coastal fishes with small mouths
and front teeth very like our own. The
Sheepshead hunts for shellfish in shallow,
muddy water from Nova Scotia through the
Gulf of Mexico and can be caught from
urban piers and bridges along the way.

PINFISH   To 14 in.
This member of the porgy family is identified
by a *dark smudge* on its "shoulder." It has a
reputation as a notorious bait stealer at
coastal fishing spots from Rhode Island
through the Gulf of Mexico.

SHINER SURFPERCH   To 7 in.
Surfperches are West Coast fishes, but this
one is no surfer. It prefers calm waters and
swarms into bays and estuaries, where it is
easily caught with small hooks from piers. It
is silvery and has a *turned-up nose.*

STARRY FLOUNDER   To 3 ft.
Flatfishes have pancake-shaped bodies,
with both eyes on one side. Unmistakable
with its boldly barred, *orange and black fins,*
the Starry Flounder can be caught from
piers and bridges from Alaska to Santa Bar-
bara. Flatfish are born, like other fish, with
eyes on both sides of their heads. As the fish
matures and becomes a bottom-dweller, one
eye migrates to the other side of its head.
Most flatfish species are either right-eyed,
with eyes on the right side of the head, or
left-eyed. The Starry Flounder, however, is
"ambidextrous"—its "top" can be either
side.

WINTER FLOUNDER   To 2 ft.
In wintertime, this right-eyed flounder with
a *small mouth* comes close to shore and is
the flatfish most commonly caught from
bridges and piers from Labrador to the
Chesapeake Bay. Its color is variable.

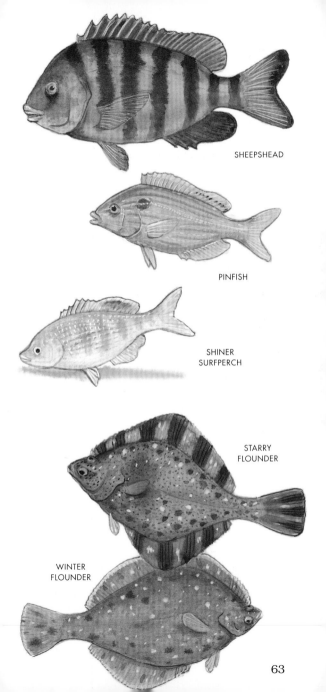

SHEEPSHEAD

PINFISH

SHINER
SURFPERCH

STARRY
FLOUNDER

WINTER
FLOUNDER

63

Amphibians

The amphibians—salamanders, frogs, and toads—need moist environments, so they are usually found close to water.

SPOTTED SALAMANDER   8 in.

Salamanders are sensitive to pollution because of their thin-skinned bodies and eggs, so few are found in urban areas. One exception is the striking *yellow and black* Spotted Salamander of eastern North America. It survives by living underground in woodlands near temporary spring ponds. It emerges in very early spring to breed and lay eggs.

REDBACK SALAMANDER   5 in.

Lungless salamanders, like the Redback, breathe through their skins. They are surprisingly widespread across North America, living in damp, untrampled leaf litter. They lay their moist eggs on land and then shield them with their slender bodies to keep them from drying out. There are more than you might think in the nearby woodlot.

BULLFROG   8 in.

The often noisy frogs and toads, with their no-neck look, muscular hind legs, and conspicuous eardrums, are familiar to almost everyone. Frogs return to water to breed. The absence of frogs and tadpoles from fresh water is an ominous sign of pollution. The widespread, deep-voiced, greenish Bullfrog is our *largest* frog and can be found in urban areas. It says *jug-o-rum*.

GREEN FROG   4 in.

Found east of the Mississippi, the Green Frog looks like a small Bullfrog but has *ridges* running from its eyes along its sides. It gives a short yelp of surprise as it leaps for the safety of the water and does quite well in urban ponds and ditches.

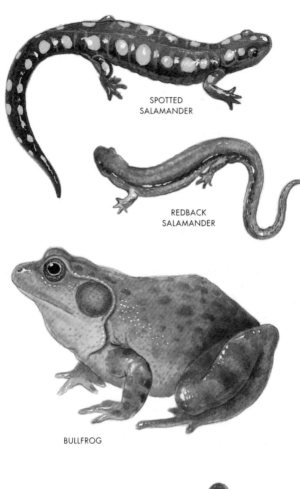

SPOTTED
SALAMANDER

REDBACK
SALAMANDER

BULLFROG

GREEN
FROG

LEOPARD FROG — 5 in.

The handsome, distinctly *spotted* Leopard Frog is slender with a pointed nose and has light-colored folds of skin running from eye to hind legs. It is an eastern frog that is often found hunting among grasses far from water.

RED-LEGGED FROG — 5 in.

Look for the Red-legged Frog all along the West Coast, in damp woods and near ponds. Golden to grayish brown, with flecks of black and a *tinge of pink* beneath its hind legs, this frog calls with a creaky voice.

GRAY TREEFROG — 2 in.
PACIFIC TREEFROG — 2 in.

The nighttime trilling calls of treefrogs are sometimes mistaken for those of insects. The single-noted, melodic warble from the trees in the East is the Gray Treefrog. Along the West Coast, the two-noted, high, peeping call belongs to the Pacific Treefrog. The large, adhesive toe pads of these frogs help them climb trees and walls in developed areas where there is moisture and cover.

WOODHOUSE'S TOAD — 5 in.

As if to make up for their warty appearance and the poison-seeping glands on their necks, many toads sing beautifully, giving a long, low trill from their breeding ponds and puddles. Toads have drier skins than frogs and can be found farther from water, often in our yards and gardens. The widespread Woodhouse's Toad sometimes feeds on insects drawn to outdoor lights. Secretions from the toad's glands do not cause warts, but they can irritate skin or eyes. If a predator can get past the bad taste of a toad, it can be poisoned.

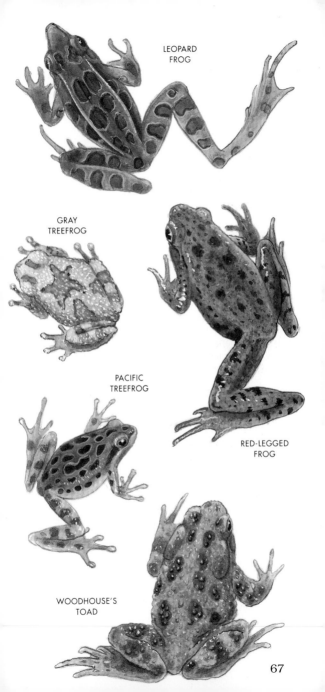

LEOPARD
FROG

GRAY
TREEFROG

PACIFIC
TREEFROG

RED-LEGGED
FROG

WOODHOUSE'S
TOAD

67

Reptiles

The reptiles include crocodiles, turtles, lizards, and snakes. Reptiles are generally shy of people and not very numerous, but a few species, such as garter snakes, are extremely widespread and abundant even in developed areas.

AMERICAN ALLIGATOR　　　　**To 18 ft.**

Formerly hunted almost to extinction for its skin, the American Alligator is now making a strong comeback from coastal North Carolina south to Texas. Where the territories of people and alligators overlap, the reptile adds an occasional pet to its normal diet of fishes, other reptiles, small mammals, and birds. The snout of an alligator is *much broader* than that of a crocodile.

SNAPPING TURTLE　　　　**To 30 in.**

The primeval looking Snapping Turtle, with its large head, powerful jaws, and jaggedly keeled tail, is widespread over muddy freshwater bottoms east of the Rocky Mountains. Out of the water, snappers can strike swiftly and have a powerful bite, but they are usually docile when in the water and will withdraw even if stepped on. Female snappers are often seen when they leave the water to lay their eggs, with a single-mindedness that makes them oblivious to human presence. It is not uncommon to find the blackish, long-tailed hatchlings scrambling around sandy edges of parking lots near water.

MUD TURTLE　　　　**6 in.**
MUSK TURTLE　　　　**6 in.**

These turtles are often called "stinkpots" because of the repellent smell produced by glands under their shells. Both have *high, domed shells*, and the Musk Turtle has *two light stripes* on its head. Stinkpots fare well even in polluted waters because they feed on dead fish.

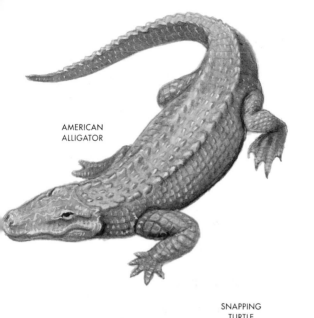

AMERICAN
ALLIGATOR

SNAPPING
TURTLE

MUD
TURTLE

MUSK
TURTLE

69

RED-EARED SLIDER — To 16 in.

Sliders are the turtles that were once sold as babies in pet stores. They have *marbled yellow and dark patterns* on their backs and necks that darken with age. They are seen throughout the South basking in the sun on rocks or logs, often piled on top of one another. Eating a varied diet that includes scavenged food, they will follow waterways into developed areas if conditions are suitable for egg laying.

PAINTED TURTLE — To 9 in.

Painted Turtles are widespread in the East and upper West.They are alert and feisty black-shelled pond turtles with attractive *red and black patterns* on the edges of their shells. Omnivorous eaters and scavengers, they can live fairly well in urban areas.

GULF COAST SOFTSHELL — To 10 in.

The *snorkel-nosed* softshell turtles are well described as "animated pancakes." They are thoroughly aquatic, fast-swimming, *flat-bodied* turtles that have *leathery skins* instead of hard plates on their shells. Should you be fast enough to catch one, beware: very long necks and sharp-edged jaws make these turtles dangerous biters.

RED-EARED
SLIDER

PAINTED
TURTLE

GULF
COAST
SOFTSHELL

71

Of the many lizards in warmer parts of North America, a few may be seen in urban areas. Lizards are distinguishable from salamanders (see page 64) by the claws on their toes and their scaled skin. Many have tails that break off easily (and eventually regrow) if the lizard is nabbed by a predator.

MEDITERRANEAN GECKO   5 in.

This chirping gecko was introduced to Florida in the 1930s and has spread throughout the Gulf region. Nocturnal and common near buildings, it can be seen clinging to walls and ceilings with its claws and the minute suction cups on its toes.

GREEN ANOLE   To 8 in.

The Green Anole can change color from green to brown to match its surroundings. The anole may be seen displaying its *pink throat fan* or doing push-ups from the walls, shrubs, and palm fronds where it lives. It is found throughout the Gulf region.

TEXAS HORNED LIZARD   To 7 in.

Short-tailed, stout, and *prickly,* horned lizards (often mistakenly called "horned toads") are well known to children in the West. In self-defense, the horned lizard can burst small blood vessels surrounding a special sac near its eyes and shoot a distracting jet of blood at its pursuers.

FENCE LIZARD   To 9 in.

The blotchy and brownish Fence Lizard is found throughout the South and Southwest. It displays from stumps and posts, bobbing up and down and flashing the *blue patches* on its sides and belly.

WESTERN SKINK   3 in.

Skinks are streamlined, alert, and active lizards. They often live in debris near buildings, usually near moisture. Skinks are very often *striped* from head to tail. Young ones have brilliant *blue tails.*

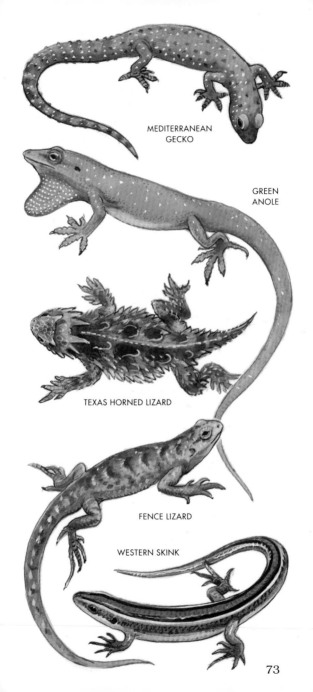

MEDITERRANEAN
GECKO

GREEN
ANOLE

TEXAS HORNED LIZARD

FENCE LIZARD

WESTERN SKINK

73

Many snakes are found in developed areas, and a few thrive even in urban settings. Snakes use their tongues to collect odor particles from the air. All snakes are carnivorous and swallow their prey whole.

EASTERN HOGNOSE SNAKE To 3¾ ft.

This nonpoisonous, blotched or dusky snake with its *upturned nose* is found east of the Rockies. It does its best to look dangerous by opening its mouth wide, spreading its neck in a menacing hood, and tightly coiling its tail in a good imitation of a rattlesnake. If these ploys fail, it rolls over and plays dead.

MILK SNAKE To 4 ft.

There are advantages to looking dangerous even if you are a harmless snake simply hanging out under trash and logs, hunting other reptiles and rodents. The widespread *reddish, yellow, and black banded* Milk Snake resembles the truly dangerous coral snake (see page 76).

NORTHERN WATER SNAKE To 4 ft.

From the Mississippi basin eastward, aggressive water snakes follow rivers and ponds into developed areas. They are active night and day, hunting frogs, small fish, and rodents. The muted, vaguely *diamond-shaped* markings look a little like those of poisonous diamondback rattlesnakes and Cottonmouths. Water snakes are not poisonous, but they can give a nasty bite.

PLAINS GARTER SNAKE To 3½ ft.

Reassuringly *striped from end to end* in black and variously flecked yellow, the widespread and familiar garter snakes resemble no poisonous snakes in North America. Semiaquatic, they often follow waterways into urban areas. If caught, they will release a foul-smelling musk and may try to bite—but they quickly calm down.

EASTERN
HOGNOSE
SNAKE

MILK
SNAKE

NORTHERN
WATER
SNAKE

PLAINS
GARTER
SNAKE

75

Although 92 percent of North American snakes are harmless, a few are dangerous. Remember, though, that venomous snakes are trying as hard to avoid a confrontation as you are. Just watch where you put your hands and feet if you are in snake country.

CORAL SNAKE   To 3 ft.

Coral Snakes have powerful venom that affects the central nervous system. *Red, yellow, and black banded,* these snakes of the coastal south and far Southwest resemble many harmless snakes, like the Milk Snake. Various rhymes help to correctly identify this snake, but none is completely accurate. Remember "black head, you're dead" and "red touches yellow, kill a fellow." Although a few harmless snakes have one or the other of these characteristics, you're wise to keep your distance from any black-headed snake whose red and yellow bands touch.

COPPERHEAD   To 4 ft.

Bulging cheeks and *diamond-shaped* markings identify members of the poisonous pit viper family. The Copperhead, found in the southeastern United States, has a reddish tinge; its "diamonds" are the pale markings rather than the darker ones.

COTTONMOUTH   To 6 ft.

Another pit viper, the Cottonmouth, also called the water moccasin, lives in waterways, which it might follow into urban areas in the Southeast. Aggressive and very poisonous, this murky-colored snake warns by opening its mouth wide, showing the white inside. It swims with its head lifted well above the water.

EASTERN DIAMONDBACK   To 8 ft.
RATTLESNAKE

Our largest and most dangerous snake, this rattler has a large *triangular* head, *diamond* patterns and a buzzing tail rattle. Watch for it in deserts, rocky areas, piney flatlands, and abandoned buildings.

CORAL
SNAKE

COPPERHEAD

COTTONMOUTH

EASTERN
DIAMONDBACK
RATTLESNAKE

77

Birds

Migrating birds pass over and through our cities, but many species have adapted to and even thrive on the conditions that cities offer.

BROWN PELICAN   3$\frac{1}{2}$ ft.

The fish-eating Brown Pelican, a skilled diver, is a large, brown and white bird with a long, sturdy bill and *enormous throat pouch*. It shares our sea walls and piers along the East and Gulf coasts.

DOUBLE-CRESTED CORMORANT   33 in.

The cormorant is an angular *black* bird that dives from the surface after fish. Its feathers are not waterproof, which helps the bird swim faster underwater, but the feathers do get wet. That's why the cormorant is often seen standing on buoys and pilings with its wings "hung out to dry."

GREAT BLUE HERON   4$\frac{1}{2}$ ft.
GREAT EGRET   3 ft.

Many herons and egrets were hunted almost to extinction for their plumes, which were fashionable on women's hats in the 1800s. The widespread, *very tall, bluish* Great Blue Heron follows waterways into urban areas in search of fish and frogs. The black-legged, white Great Egret breeds in the South but wanders north each summer.

CATTLE EGRET   18 in.

The *chunky, buff-headed* Cattle Egret is familiar from nature films of Africa, where it rides on the backs of large grazing animals that stir up small creatures for it to eat. Introduced to South America, it is spreading north rapidly, following plows in the fields and mowers along our highways instead of water buffalos and zebras.

SNOWY EGRET   25 in.

The delicate Snowy Egret is common in salt marshes in the East and along the coast and in inland ponds in the West. It has *bright yellow feet* and beautiful plumage, especially during breeding season.

BROWN
PELICAN

DOUBLE-CRESTED
CORMORANT

GREAT
EGRET

GREAT BLUE
HERON

CATTLE
EGRET

SNOWY
EGRET

79

MUTE SWAN   5 ft.

The *long necks* of the heavy-bodied Swans help them gather food from the bottoms of ponds and estuaries. The majestic Mute Swan, partially domesticated since the Middle Ages, is somewhat tolerant of people. Introduced into North America around 1900 to grace the estates of the rich, Mute Swans have taken over many environmentally important brackish ponds from Rhode Island to Maryland and at various locations across the continent. Though they are indisputably elegant birds, with their S-curved necks and arched wing posture, swans are wasteful feeders and often drive away or kill every other bird species at the ponds they inhabit.

CANADA GOOSE $2^1/_2$–$5^1/_2$ ft.

The sociable Canada Goose, with its *black head and neck* and *white cheeks*, comes in two models: the nonmigrating "giant" race (68 in.) and the smaller, migrating race (30 in.). Both fly in V formations. Normally gleaners of open fields and grassland, Canada geese find our urban parks, corporate lawns, median strips, and playgrounds very much to their liking. The numbers of the "giant" Canada have increased so much that they have become a problem in some areas. A flock of these birds, each of which produces almost a pound of droppings per day, can make a real mess.

MALLARD   25 in.

The familiar male Mallard, with his *iridescent green head* and *white neck ring,* is often seen with his pleasantly camouflaged partner and their ducklings. The Mallard is the largest of our dabbling ducks, those that go bottoms-up in search of food in shallow water. They are the ancestors of the domestic white duck and are found all over North America, including city ponds and estuaries.

MUTE
SWAN

CANADA
GOOSE

giant
race

migrating race

female

male

MALLARD

81

TURKEY VULTURE   30 in.

The *bare-headed* Turkey Vulture is a soaring, black bird with *spreading feathers* at the tips of its large wings. Like other flesh-eating birds, it has a hooked beak that helps it cut through the skin, sinews, and bones of its prey. Road-killed animals draw the Turkey Vulture near our highways.

PEREGRINE FALCON   20 in.

The Peregrine has been the hunting falcon of choice in the Middle East for centuries. With its pointed wings and long tail, the Peregrine is designed for stunning speed. It can "stoop" (dive) after birds at speeds above 100 m.p.h. The numbers of this and other bird-eating predators dwindled in the 1940s and 1950s as pesticides eaten by their prey accumulated in their bodies. Many species were nearly extinct by the early 1970s. Heroic efforts saved the elegant, *slate gray, black-cheeked* Peregrine Falcon. It does well in a few of our cities, nesting on skyscraper ledges and preying on birds that are free from farm pesticides.

AMERICAN KESTREL   11 in.

The Peregrine's tiny country cousin, the American Kestrel, has been lured near urban areas by mowed highway edges and open train yards, where it can be seen sitting on wires looking for insects, small birds, and rodents. It has *white cheeks,* a *reddish back* and slate blue wings.

RING-NECKED PHEASANT   34 in.

Although it looks extravagant, with its bright colors and *long, barred tail,* the Ring-necked Pheasant has rather chickenlike habits and forages on the ground for food. The pheasant originated in Asia and was introduced into North America, where it is now widespread. Most common in open grassland, the pheasant has adapted well to weedy fringes of developed areas. If you surprise one in hiding, it will give its loud *kuk-kuk* call and noisily take to the air.

TURKEY
VULTURE

PEREGRINE
FALCON

AMERICAN
KESTREL

RING-NECKED
PHEASANT

83

KILLDEER   10 in.

The insect-eating Killdeer has adapted well to airports, golf courses, dumps, and train yards. It has *two bold bands* across its white chest and a repeated and anxious *kildea-kildea-kildea* call. This ground nesting bird is also known for its broken-wing act, with which it lures intruders away from its young.

HERRING GULL   25 in.

Some of the sturdy, plentiful, scavenging gull species were once hunted almost to extinction for their feathers and eggs. Protected now, they thrive on our piles of refuse. The large, *gray-backed* Herring Gull, with its *pinky-gray legs* and the *red spot* on its lower bill tip, is found along all North American coasts and well inland. Savvy, long-lived, and opportunistic, some gulls have learned to hang about public places on the waterfront or at dumps, in hopes of stealing some food.

CALIFORNIA GULL   17 in.

The smaller, *yellow-legged* California Gull is found along the West Coast, nesting inland on the prairies in summer. It is probably the only bird to have a monument erected in its honor: These gulls consumed the locusts that were devastating the first crops of the Mormon pioneers in Utah.

RING-BILLED GULL   17 in.

The widely distributed Ring-billed Gull, with a *black ring* circling the end of its bill, is smaller than the Herring Gull. It seeks food inland, following plows and hanging around parking lot dumpsters, hoping for the odd bit of food.

LAUGHING GULL   16 in.

The *black-headed* Laughing Gull is named for its piercing *ha-ha-ha-haaaa* cry. Found along the Atlantic and Gulf coasts, it can become a cheeky picnic thief at public beaches.

KILLDEER

HERRING
GULL

CALIFORNIA
GULL

RING-BILLED
GULL

LAUGHING
GULL

85

ROCK DOVE   13 in.

This is the ubiquitous pigeon that thrives in
our cities and towns, nesting on building
ledges and eating just about anything it can
find. Usually gray, with *barred wings* and
pink legs, Rock Doves have lived near
humans for 5,000 years. Pigeons and doves
feed their young with a unique "pigeon milk"
consisting of half-digested seeds and fluids
secreted from glands in their throats. The
homing abilities of pigeons are famous. Our
most sophisticated mechanical navigation
systems are cumbersome compared with
those in the brains of these small-headed
birds. Exactly how they do it is still a mystery, but in addition to their excellent
senses of sight and smell, Rock Doves can
hear long-distance, low-frequency sounds,
see ultraviolet light, and sense the Earth's
magnetic field.

MOURNING DOVE   12 in.

This graceful, gently cooing dove, with its
soft gray-brown coloring and *long, tapered,
white-edged tail,* is found throughout North
America. A native, ground-feeding, seed-
eating bird, it has benefited from our forest-
clearing activities and thrives in city parks.

RINGED TURTLE DOVE   12 in.

Imported as pets, escaped Ringed Turtle
Doves have become well established in
warm cities such as Los Angeles and Miami,
and are doing well in a few places as far
north as Baltimore. This slender, beige,
endlessly cooing dove has a semicircular
black collar.

MONK PARAKEET   10 in.

Escaped pet parrots sometimes survive
around developed areas and establish local
populations. Flocks of communally nesting,
noisy, green, *gray-crowned* Monk Parakeets
have become locally established in North
America. The Monk is one of some 10 species of escaped parrots that may take you by
surprise in your neighborhood.

ROCK DOVE
"pigeon"

MOURNING
DOVE

RINGED
TURTLE DOVE

MONK
PARAKEET

87

SNOWY OWL   20 in.

Owls are generally nocturnal birds of prey with sharp talons and hooked beaks. An owl cannot move its large, light-gathering eyes in their sockets, so it must turn its head in order to see. A truly exotic urban bird, the large, ground-dwelling white Snowy Owl breeds in the far North but travels south as far as the middle United States in winter. It may be seen hunting rodents at urban airports, which it considers an acceptable substitute for the treeless tundra.

BURROWING OWL   9 in.

This little, brown, white-spotted, ground-dwelling owl has a continuous *white eyebrow*, *long* white legs, and a disgruntled expression. Perched near its hole, it hunts rodents in the prairielike expanses of airports in western North America.

COMMON NIGHTHAWK   9 in.

The Nighthawk is a slim, night-flying bird with *whiskers*, pointed wings, and a *white bar* across the underside of each wing tip. It has a tiny bill but an enormous mouth, and can be seen flying about street lights with mouth agape, scooping up flying insects. The Nighthawk is the city cousin of the Whip-poor-will, and like that bird is more often heard than seen, its jarring *peent* call rising above the noise of the city. A ground nester, the Nighthawk satisfies its instincts in the city by nesting on flat, gravelly rooftops.

CHIMNEY SWIFT   6 in.

Insect-eating Chimney Swifts are bullet-shaped, dark-colored, fast-flying birds that appear tailless. From a city roof at sunset you can watch their swirling, excited gatherings before they funnel down into the chimneys and air shafts where they roost.

BURROWING OWL

SNOWY OWL

COMMON NIGHTHAWK

CHIMNEY SWIFT

nest

89

RUBY-THROATED HUMMINGBIRD   3 in.
RUFOUS HUMMINGBIRD   3 in.

The Rubythroat is found in the East, the Rufous in the West. Both kinds of hummers have beautiful iridescent plumage—*green* for the Rubythroat, *red-brown* for the Rufous—and the males have brightly colored throats. Females of both species are more modestly colored. Watch for these nectar feeders hovering with invisible, humming wingbeats in front of red flowers, even in developed areas.

DOWNY WOODPECKER   5 in.

North American woodpeckers have strong, sharp bills for excavating nest holes in dead wood and digging in bark for insects. This small bird with a *black and red cap* is comfortable around people and is the most likely of the woodpeckers to visit urban trees and suet feeders.

BLUE JAY   12 in.

This bold *blue* bird with its rakish *head crest* and its *jay! jay!* call tolerates the presence of people throughout the East and upper Northwest. Though jays are famous for their raucous conversations with one another and for loudly broadcasting the approach of predators, they are impressively silent while raising their young. The Blue Jay eats mostly nuts, seeds, and insects, but it has been known to take an egg or nestling from time to time.

AMERICAN CROW   20 in.

The *big, black* crow travels as far as 50 miles morning and evening to its communal roost. Crows are quite intelligent birds, and will exploit all food sources. Crows are often heard scouting around early in the morning; later they go to fields, suburbs, and along highways for serious feeding. Some 23 distinct calls of these remarkably vocal birds have been identified. The meaning of the calls is still a mystery.

RUBY-THROATED
HUMMINGBIRD

RUFOUS
HUMMINGBIRD

DOWNY
WOODPECKER

BLUE JAY

AMERICAN
CROW

91

BLACK-CAPPED CHICKADEE   4 in.

The widespread chickadee, with its black cap and chin, nests secretively in northern woodlands and migrates to warmer areas in winter. It is a tame little bird and a lively presence at feeders. Listen for its bright *chick-a-dee-dee-dee* call. In early spring the chickadee makes a plaintive *fee-beee* call.

HOUSE WREN   4 in.

Distinctive because of its *upward tilting tail*, the energetic little brown wren is known for nesting in odd places: pockets of forgotten coats, old gloves, or other cavities. We can hear its penetrating, repeated bursts of song in backyards and city parks.

AMERICAN ROBIN   9 in.

The *brick red-breasted* robin is a melodious singer, calling *cheer up, cheerilee, cheer up* from trees and shrubs. The robin hunts worms and insects in summer, running a few steps and cocking its head to look for small movements. In winter, robins eat berries in woodlands. Robins become fiercely territorial when they are nesting. In their eagerness to evict other males, robin husbands sometimes "see red" and attack other small red objects—like balls or hats—in their territories.

NORTHERN MOCKINGBIRD   10 in.

With what looks like a leap of pure joy, the male mockingbird springs into the air from a TV antenna or rooftop, singing exuberantly. This virtuoso mimic can sing the songs of up to 180 birds in addition to his own. He may sing all night long on spring moonlit (or streetlamp-lit) nights, in an attempt to impress potential mates. Originally from the South and Southwest, these *gray and white, long-tailed* birds have adapted well to developed areas and spread widely northward, eating insects and berries from our ornamental plants and bestowing bird song where there might otherwise be none at all.

BLACK-CAPPED
CHICKADEE

HOUSE
WREN

AMERICAN
ROBIN

NORTHERN
MOCKINGBIRD

93

EUROPEAN STARLING   8 in.

Kind things have been said about the European Starling—but most of them have been said in former centuries. Nostalgic immigrants introduced starlings to North America, where they aggressively displaced native birds of similar habits. Since then, starlings have multiplied rapidly, doing millions of dollars of crop damage every year. Starlings are cavity nesters, and readily nest in niches in buildings. The male lures potential mates by building a rudimentary nest; if a female accepts him, she throws his nest out and builds another. Mated males, bachelor and immature birds—and all starlings outside of nesting season—spend their nights in huge, noisy, dropping-spattered communal roosts in groves or under bridges. In breeding season, starlings are *iridescent black* birds with long, *yellow bills* and short tails. Outside of breeding season, their plumage is dusky brown flecked with white. Their song repertoire includes a variety of clicks, whistles, and rattles.

SONG SPARROW   6 in.

Like other birds that eat seeds, such as finches and cardinals, sparrows have short, sturdy beaks. If the sparrow in the shrubbery has a strongly *striped breast* and calls *maids maids maids put on your tea kettle ettle ettle* it is probably a Song Sparrow. If it dips its tail in flight, you can be sure of it.

CHIPPING SPARROW   5 in.

This modest little sparrow with a *rusty cap, dark bill,* and *gray rump* is found in parks and playgrounds everywhere. Its song is a reedy trill.

NORTHERN JUNCO   6 in.

Juncos look different in different parts of the country, but they interbreed freely and so are considered one species. Generally dark sparrows with *black hoods* and *white undersides,* they visit city parks and feeders in winter.

EUROPEAN
STARLING

SONG
SPARROW

CHIPPING
SPARROW

NORTHERN
JUNCO

95

HOUSE FINCH   5 in.

Originally from the Southwest, this mild
and adaptable finch has spread with the
help of feeders throughout most of the
United States and into British Columbia.
The male, with his faded *red crown* and
breast, was once kept as a pet for his pleas-
ant song. Now, the House Finch's soft,
questioning *cheep?* and its bubbly, cheerful
song emerge from ledges, vines, trees, and
shrubbery. It nests readily even in downtown areas, taking advantage of bits of
greenery such as hanging plants and window boxes.

HOUSE SPARROW   6 in.

Many native birds fled from the widespread
habitat destruction caused by the growing
towns and cities of North America. In the
absence of the birds, insects flourished.
Immigrants remembered the insect- and
seed-eating "sparrows" (actually weaver
finches) of crowded, long-urbanized Europe
and imported them. Unfazed by the presence of people, the stubby, gray-capped,
black-bibbed males and the grayish females
prospered and spread throughout urbanized North America, filling the empty niches
left by native birds and aggressively driving
out the rest. Argumentative and social,
House Sparrows forage in small flocks,
enjoy dust baths in dry spots, and negotiate
chain link fences with skill, squabbling or
cheeping loudly and repeatedly.

RED-WINGED BLACKBIRD   7 in.
YELLOW-HEADED BLACKBIRD   9 in.

Both the more eastern Red-winged and the
western Yellow-headed blackbirds will follow reedy marshes into urbanized areas.
Where they overlap, redwings choose the
drier parts of the marsh. The redwing's joy-
ful *bubbleee* call is the first bird song heard
in early spring. The handsome yellowhead's
song is less appealing; he has a call that
sounds like a creaking hinge.

HOUSE FINCH

male

female

HOUSE SPARROW

male

female

YELLOW-HEADED BLACKBIRD
male

RED-WINGED BLACKBIRD
male

97

Mammals

OPOSSUM   To 35 in.

This is the only marsupial (mammal with a pouch) in North America. It is famous for playing dead—"playing possum"—when threatened. It has a *long nose*, a ghostly *white face* and *naked, prehensile tail*. The solitary Opossum has wandered northward from its original southern home, eating roadkill, berries, insects, and frogs. It hides in brushy margins of waterways, emerging at night to forage. Young marsupial mammals are extremely tiny and undeveloped when they are born. Their mothers carry them in a pouch while they finish growing. The babies often ride on their mother's back after they grow too large for her pouch.

SHORTTAIL SHREW   To 5 in.

Not every little rodent the cat brings in is a mouse. This fierce little hunter is extremely common, but seldom visible, in the eastern half of North America. Like moles, shrews are almost blind and rely instead on an excellent sense of smell and touch. The Shorttail Shrew rushes through hidden runways under leaves, grasses, snow, and house foundations, hunting insects and other small animals continuously. It will also fearlessly attack larger prey, such as snakes, and kill them with the aid of its poisonous saliva. You and your cat are too large to be dangerously affected by a shrew's venom.

EASTERN MOLE   To 8 in.

Immensely strong for its size, the velvety mole lives its whole life underground. It can dig tunnels at the rate of 12 feet per hour with its wide, *shovellike forepaws*. Its tiny eyes (unnecessary for life underground) are covered with skin. It locates worms and grubs with its keen hearing and sensitive whiskers. The Eastern Mole lives east of the Rockies, and it might raise some tunnels in your backyard.

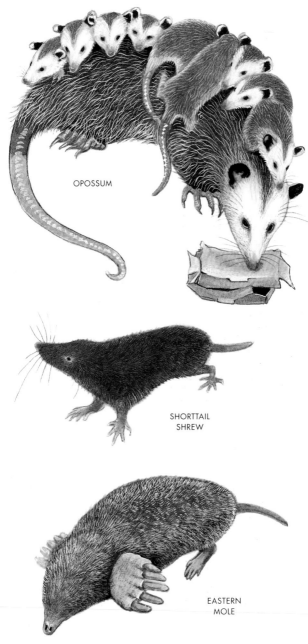

OPOSSUM

SHORTTAIL
SHREW

EASTERN
MOLE

99

LITTLE BROWN MYOTIS
CALIFORNIA MYOTIS   To 3 in.

Bats are the only true flying mammals. Their wings are formed by skin that stretches around the long bones of their hands and to their tails. Although they look like mice, they are actually more closely related to shrews. "Evening bats" like these two are common across North America. These bats are beneficial animals, as they eat huge numbers of insects such as mosquitoes. Like other insect-eating bats, they locate their prey in the dark by using sound waves. In summer, these bats may roost during the day in barns, under bridges, or in crevices in our buildings.

BIG BROWN BAT   To 5 in.

This larger version of the little myotis bat is the bat that most commonly finds its way into houses throughout North America. Fond of roosting in storm sewers or attics, it is the bat you are most likely to meet personally. If you find a bat inside the house, close the door of the room where you found it and leave a window open; it will fly out at dusk.

MEXICAN FREETAIL BAT   To 4 in.

This southwestern bat is famous for roosting in huge numbers in caves or under bridges, even in cities. Freetail bats have *"free tails"*—they are not enclosed in the skin membranes that extend between their legs.

ARMADILLO   To 3 ft.

This odd but adaptable burrowing mammal is covered with *horny plates,* with *jointed bands* in the middle for flexibility. The female always gives birth to identical quadruplets, all four of the same sex. While searching along roads for carrion, Armadillos are often killed by cars themselves.

CALIFORNIA MYOTIS

LITTLE BROWN MYOTIS

BIG BROWN BAT

MEXICAN FREETAIL BAT

ARMADILLO

101

EASTERN COTTONTAIL   To 18 in.

Widespread east of the Rockies, this rabbit, with its *white "cotton ball" tail*, likes what humans have done to the land. We have created more edges—grassy areas with brushy cover nearby, like roadsides, cemeteries, backyards, and parks—where it can graze. Rabbits have shorter legs than hares, such as the Jackrabbit, and give birth to naked and dependent young.

WHITETAIL JACKRABBIT
BLACKTAIL JACKRABBIT   To 24 in.

Between the two of them, the *long-legged, long-eared* Whitetail and Blacktail jackrabbits cover the American West. In their view, grassy airports make reasonable homes, and they are often seen feeding on grass beside highways. Jackrabbits are famous for their speed and have been clocked at 40 m.p.h. Unlike rabbits, hares are born fully furred and able to move about.

EASTERN CHIPMUNK   6 to 10 in.

There are many species of this pert, ground-living squirrel throughout our region. All have *facial stripes* and *black-bordered white lines* down their sides. Chipmunks are so well camouflaged in dry leaves that they are virtually invisible unless they move—which they do a lot. The Eastern Chipmunk can tolerate urbanized woodlands, where it lives on nuts, seeds, and small invertebrates. Like many squirrels, chipmunks hibernate, sleeping deeply during winter months.

WOODCHUCK   To 24 in.

Also known as the groundhog, this chunky, plain grayish brown marmot always sleeps through Groundhog Day into March or April. An edge specialist, this large member of the squirrel family is familiar to highway travelers in the Northeast westward through the northern prairies to Alaska.

EASTERN
COTTONTAIL

WHITETAIL
JACKRABBIT

BLACKTAIL
JACKRABBIT

EASTERN
CHIPMUNK

WOODCHUCK

103

13-LINED GROUND SQUIRREL   To 14 in.

White stars and stripes distinguish this lovely little red-brown ground squirrel. It has many shy country cousins, but this animal is attracted to golf courses and mowed roadsides from the Texas Gulf coast north to the Canadian prairies.

GRAY SQUIRREL   To 26 in.

Taking advantage of our bird feeders, our trash, our briefly abandoned chocolate bars, and our shade trees—thoughtfully linked by utility wires—the opportunistic Gray Squirrel is familiar to everyone in the eastern United States and some places farther west. Its big, *bushy tail* serves as a balance, a blanket, and a warning flag.

WHITE-FOOTED MOUSE   To 9 in.

There are more than 1,000 species of rats and mice in the world. Only a few of these have important bad effects on humans (see page 106), and the eastern White-footed Mouse is not usually one of them. It is a "country mouse" who only occasionally drops by our houses for a visit during cold weather. It resembles many of its wild cousins throughout North America.

MEADOW VOLE   To 7 in.

Widespread throughout all our cooler regions, this *short-tailed, blunt-nosed* field mouse has been called a machine for turning grass into meat. Living outside anywhere there is grassy cover, it often falls prey to our cats.

MUSKRAT   To 30 in.

This large aquatic vole, with its *long, nearly hairless tail* that helps it swim, follows waterways into urban areas throughout North America. Feeding on water plants, frogs, fish and young birds, it builds lodges using plant materials and mud, instead of wood as a beaver does. About 8 million muskrats are trapped each year for their waterproof, shiny, dense pelts.

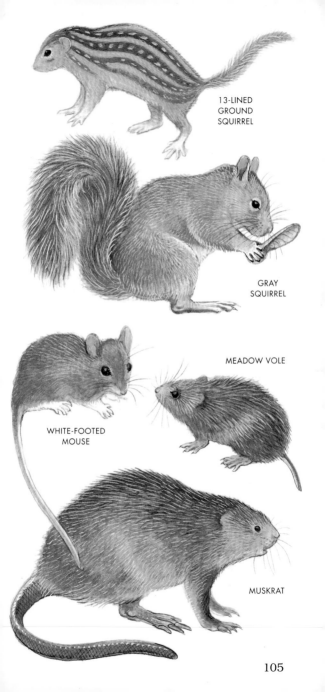

13-LINED GROUND SQUIRREL

GRAY SQUIRREL

MEADOW VOLE

WHITE-FOOTED MOUSE

MUSKRAT

105

BLACK RAT   To 20 in.

The Black Rat is variably colored. Known also as the ship rat or roof rat, this world citizen climbs with skill, running along utility wires and ship ropes and living in attics. This destructive rat is most often found in heavily developed coastal areas. Like its scaly-tailed cousins, the Norway Rat and the House Mouse, these animals have lived and traveled with us—in our houses, our granaries, our ships, and our saddlebags— for all of recorded history. To distinguish a Black Rat from a Norway Rat, look for its slightly *sharper nose*, slightly *larger eyes and ears* and its tail, which is a bit *longer than its body*. This rat brought us the flea, which brought us the bacteria that caused the horrifying epidemics of Black Death in Europe beginning in the 13th century. It was later joined in this unintentional deadly work by the Norway Rat. When the two species meet, the milder mannered Black Rat retreats—or moves upstairs.

NORWAY RAT   To 22 in.

This large, *blunt-nosed, small-eared* rat is also found all over the world. Known as the sewer rat, wharf rat, or common rat, it can swim and dive. A ground-loving rat, it digs elaborate underground burrows, with rooms for storage, nurseries, bathrooms, and many exits. It has fared well on farms and in cities, causing extensive damage by stealing and spoiling food, spreading diseases, and, like all rodents, gnawing. Its domesticated version, however, is the white lab rat, which has given us many insights into the causes and treatments of some of our most deadly diseases.

HOUSE MOUSE   To 8 in.

This *dusky* little *long-tailed* mouse shares our food and dwellings, where it can do great damage by eating or soiling food and gnawing woodwork and electric wires. Its domesticated cousin is the familiar white lab mouse.

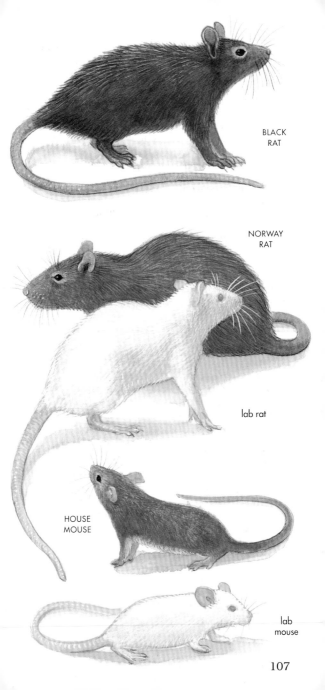

BLACK
RAT

NORWAY
RAT

lab rat

HOUSE
MOUSE

lab
mouse

107

COYOTE   To 4 ft.

Humans once destroyed all but the smartest coyotes. Then we helped them by hunting nearly to extinction their rivals the wolves, which used to keep Coyotes out of their territories. As a result, this *long-nosed, large-eared, golden brown* member of the dog family has spread eastward from its original western home with remarkable success. Tireless, intelligent, adaptable, and wary, it has found its way into numerous cities throughout North America where it eats fruit and carrion, and hunts rodents and an occasional cat.

FERAL DOG

Dogs have lived with humans for 15,000 years. Much less able than cats to fend for themselves, abandoned dogs can sometimes manage a miserable existence and even reproduce, scavenging food from trash and dumps. Such feral dogs will avoid people. Pet dogs allowed to run free, especially those that join packs of other dogs, will behave in ways that would shock their owners, including harassing and killing livestock and wild animals. Make sure your own pets are spayed or neutered, and keep them under your protective control.

RED FOX   To 3^1/$_2$ ft.

Humans have created many perfect environments for rodents, and the Red Fox has noticed. Nudged from behind by Coyotes, using culverts, overpasses, rail lines, waterways, and median strips, Red Foxes have made their way into the cemeteries, parks, and rail yards of cities across North America. There they are protected by the lack of hunters and by leash laws for dogs. But because this *red, black-footed* fox with a *white tip* on its very bushy tail is nocturnal, we rarely see it. Foxes, like many wild animals, can carry rabies. Be sure your own pets are protected with rabies shots.

COYOTE

FERAL DOG

RED
FOX

109

RINGTAIL
To 3¹/₂ ft.

In fast-growing southwestern areas where cities and suburbs overlap rocky or wooded land, you may catch a glimpse of this shy, slender, two-pound member of the raccoon family. Also known as the cacomistle, this mammal, with its *long, banded tail*, normally eats insects, small animals, plants, and fruit. It is an extraordinarily skilled climber, with sharp claws and hind feet that can rotate 180 degrees like a squirrel's. Dog food left outside, rodents, and fruit-bearing decorative trees can attract these nighttime visitors.

RACCOON
To 3¹/₂ ft.

Distinctive and clever, this nocturnal, *black-masked, ring-tailed* mammal has exploited the presence of humans. It has ranged far north and west from its original southern home. The extremely dexterous Raccoon can turn knobs, open latches, open and overturn garbage pails, pry open ventilation louvers, and brazenly use pet doors to enter houses. With its assertive hustle, rolling gait, and raised, handsome fur, it can look very formidable, and it will fight and bite fiercely if it must. Raccoons often carry the rabies virus and should never be approached.

MINK
To 28 in.

The Mink is cursed with a luxurious, *glistening, dense, brown coat;* 100 of these small, *white-throated* weasels give up their lives and pelts to make one full-length mink coat. Mink are now raised commercially for this purpose. It is a surprise to learn that this *long-bodied, short-legged* carnivore can be found everywhere in North America where there are waterways to hunt along, even in urban areas. Mink swim well, preying on fishes, frogs, muskrats, snakes, and waterfowl. Each Mink marks its personal shoreline circuit with strong-smelling musk.

RINGTAIL

RACCOON

MINK

111

STRIPED SKUNK   To 28 in.

Here is an animal with no need for camou-
flage. Its bold colors warn us of its well-
respected weapon: a potent brew of sul-
phurous alcohol and oil sprayed from
glands beneath its tail with which a skunk
can score a direct hit at 10 feet. The Striped
Skunk has a *white cap dividing into two
back stripes*. It has benefited from our presence and often feeds on garbage. Fearless
and slow, skunks cannot adapt to cars, and
many are killed on the road.

HOGNOSE SKUNK   To 27 in.

Found only in the Southwest in North
America, this *white-backed, bare-nosed*
skunk may be seen as our cities spill over
into dry, rocky areas.

SPOTTED SKUNK   To 18 in.

This playful little skunk, with its *spotted*
coat, is found in the Southeast, central
plains, and Far West. Agile and a good tree-
climber, it is less common but not unknown
in developed areas. It sometimes sprays
from a "handstand" position.

FERAL CAT

More than 30 million cats are abandoned by
their owners each year. Free-living cats live
brief, prolific, desperate, and unhealthy
lives. They kill countless numbers of song-
birds and often carry rabies.

MOUNTAIN LION   To 7 ft.

Once widespread, our only big cat has
retreated into mountainous regions of the
West, where it hunts deer and elk. Now protected, it is making a modest comeback in
urban areas that snuggle up to wilderness.
Only six human deaths have been caused
by these animals in 100 years. But in areas
where these cats are known to live, adults
should carry small children on their shoul-
ders, and joggers should avoid stooping
down. Mountain Lions fear confrontation
and avoid noise.

STRIPED
SKUNK

HOGNOSE
SKUNK

SPOTTED
SKUNK

FERAL
CAT

MOUNTAIN
LION

113

WHITETAIL DEER   To 6 ft.

The eastern Whitetail benefited greatly from the leafy, second-growth woodlands that took over abandoned farmlands. Our suburbs dislodged large predators, discouraged hunting, and encouraged dog control. In the population explosion that followed, the Whitetail has spread across all but the driest and coldest parts of North America. They are masters at remaining invisible, but you may surprise this lovely, nervous deer—the *white underside* of its flaglike tail flashing as it bounds away—almost anywhere. It can use even tiny areas such as woodsy islands next to airports and wooded places in partially industrialized areas and sprawling suburbs. Collisions kill 350,000 deer and 100 motorists each year. These deer are the most important carriers of the tiny tick that carries Lyme disease.

MULE DEER   To 6½ ft.

The Mule Deer looks much like the Whitetail, but it has *larger ears.* It also has a *black or black-tipped tail.* This relatively calm-natured deer occasionally associates casually with people in developed areas west of the Mississippi. Remember that adult bucks of all deer species are irritable and dangerous during the autumn mating season.

MOOSE   To 9 ft.

Unmistakable because of its *huge size* and *pendulous nose,* the normal range of the Moose is in second-growth woodland near water throughout Canada and in northern areas of the United States. Occasionally, a moose, usually a male, will make headlines by wandering through a downtown area, even deep in the heart of the Midwest. These animals are either young bulls in dogged search of a mate and a new territory, or they are suffering from "brain worms," nematodes that commonly afflict Moose, disorienting and eventually killing their victim.

WHITETAIL
DEER

MULE
DEER

MOOSE

115

CALIFORNIA SEA LION   To 8 ft.

Seals and sea lions are carnivorous marine mammals with all four legs modified into flippers. Sea lions have tiny but visible *ears, large front flippers,* and hind flippers that rotate forward to help them move on land. The California Sea Lion is protected by law, as are most marine mammals. Encouraged by this and by the safety of our sheltered marinas from sharks, the sleek, sociable, dark golden brown sea lions occasionally lounge about on urban docks and floats, to the delight or exasperation of nearby humans. This is the trained "seal" most often seen performing in captivity.

MANATEE   To 15 ft.

Manatees are large, slow-moving marine animals with a *single,* paddle-shaped, bone-less hind flipper. Only 2,000 of these mammals live in North America, in Florida along the Gulf Coast and up the East Coast to North Carolina in summer. They wander up warm rivers and into bayous and urban channels in their search for up to 100 pounds of aquatic plants per day. Because they like to bask near the surface, manatees are frequently injured or killed by boats.

WILD PIG   To 6 ft.

Pigs are grazing or rooting animals that have been domesticated by man for 10,000 years. Others have always lived, wary and resourceful, in the wild. Called "boars" after the males of the species, wild pigs have been introduced locally in North America for hunting. Dark-skinned, with *coarse hair* and *upturned tusks,* they have interbred with escaped domestic pigs and can be found along the West and Gulf coasts and in mountainous areas of the Southeast. They root in the ground, feeding on acorns and almost everything else. In areas of rapid human development, these wild pigs can become our neighbors. They can seriously damage fragile habitats.

CALIFORNIA
SEA LION

MANATEE

WILD PIG

117

HUMPBACK WHALE   To 50 ft.

The *long-flippered* Humpback Whale, which
migrates up and down both our coasts,
feeds by filtering small marine animals from
the water with massive, comblike horny
baleen, and breathes through the blow hole
on top of its head. Like all marine mammals, Humpbacks have exceptional hearing
and communicate using sound, which travels long distances underwater. The male
Humpback is a superb singer, with songs
lasting from 30 minutes to two hours. Both
males and females have songs about feeding. This fact was most helpful when an
adolescent Humpback affectionately known
as Humphrey swam into San Francisco Bay
and up the Sacramento River, once in 1985
and again in 1990. Recorded feeding songs,
broadcast underwater, encouraged him to
turn around and return to sea. Some
whales, for unknown reasons that may
include confusion or disease, will repeatedly
swim into shore and beach themselves.
They usually die of the pressure of their
own weight on internal organs, shock, and
overheating.

BOTTLENOSE DOLPHIN   To 12 ft.

Unlike most dolphins, which prefer the
open ocean, the playful Bottlenose comes
close to shore on both coasts. It may enter
estuaries, shallow bays, and even rivers,
where it seems to enjoy interacting with
people. Dolphins use sound waves to find
objects underwater. They are *gray above*
and *white below*. Don't confuse these mammals with the delicious game fish of the
same name.

HARBOR PORPOISE   To 6 ft.

Dusky colored, with a *dark stripe* from the
corner of its mouth to its flipper, the Harbor
Porpoise enters bays and large rivers along
the West Coast from Alaska to Baja and
from Baffin Island to Cape Hatteras along
the East Coast. Porpoises have shorter,
blunter heads than dolphins.

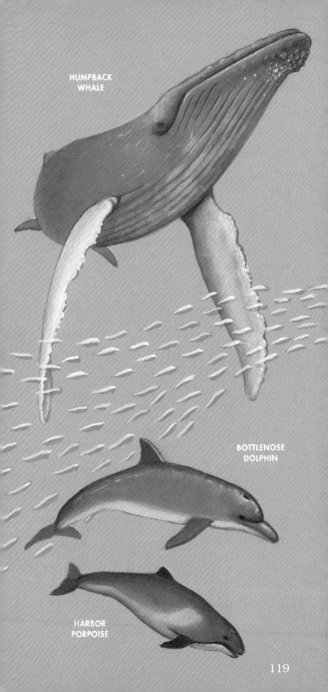

HUMPBACK
WHALE

BOTTLENOSE
DOLPHIN

HARBOR
PORPOISE

119

Plant Kingdom

Plants transform the energy of the sun directly into food. The earliest green bacteria created our oxygen-rich atmosphere; today's plants cleanse and renew it, and some can even remove toxins from the air we breathe.

CRABGRASS   To 20 in.
Crabgrass is one plant you can count on finding throughout North America, in gardens and parks and even in cracks in the sidewalk. It lives only one year, leaving its seeds behind to sprout in any crack or cranny. It has a *sprawling, crablike form*; even the seed head is rather crab-shaped.

DOWNY BROME GRASS   To 20 in.
This grass sprouts early in disturbed ground, then turns golden yellow and quickly drops its *long-awned* seeds from *loose, drooping* seed heads. It is found widely along roadsides and railroad tracks.

PHRAGMITES REED   To 12 ft.
Tall, tough, and invasive, this hardy grass has underground stems that spread quickly in moist places, sprouting in spring into dense stands of reeds with *large, plumed seed heads.* Seeds carried by vehicles and wind carry *Phragmites* to disturbed places such as ditches and swamp edges along our highways.

CATTAIL   To 8 ft.
This familiar, water-loving grass, with its unusual *sausage-shaped* flower head, likes its roots wetter than *Phragmites* does, but where the two plants jostle for position, Cattails are often forced out. Both plants help to filter and trap water pollutants, keeping them from entering our rivers and oceans.

CRABGRASS

DOWNY
BROME
GRASS

CATTAIL

PHRAGMITES
REED

121

WATER HYACINTH   To 16 in.

The floating water hyacinth, with pretty
blue-lavender flowers, was introduced from
South America and has spread rapidly,
choking rivers, streams, and ponds
throughout frost-free parts of North America. Yet this plant, with its *round waxy
leaves* and *bulbous stems*, filters pollutants
from water and provides food for the endan-
gered Manatee (see page 116).

JAPANESE KNOTWEED   To 7 ft.

This relative of rhubarb and buckwheat,
originally from Southeast Asia, has spread
all over North America, growing in dense
stands along railroads, waterways, and
roads, providing cover and seeds for animals but crowding out other plants. It has
triangular leaves and *sprays of small white
flowers* followed by *three-winged seeds*.

COMMON RAGWEED   To 5 ft.

This infamous plant thrives in poor soil,
including vacant lots and rail yards. It has
slender, silver-bottomed leaves that are
many-fingered at the plant's base and single
near its top. Its numerous tiny green flowers
shed volumes of especially irritating pollen
in late summer, causing misery for people
who suffer from "hay fever."

CHICORY   To 4 ft.

Stiff-stemmed Chicory, with its *intensely
blue* flowers in late summer, often surprises
us by popping up from crevices in city pavement. Its carrotlike root can be roasted and
used as a coffee substitute.

DANDELION   To 18 in.

An annoyance to lawn-tenders, the Dandelion is nevertheless a charming plant, with
sunny yellow flowers followed by a globe of
seeds dispersed on little gossamer parasols.
Its jagged-edged leaves are edible when
gathered early from lead-free soil.

WATER HYACINTH

JAPANESE
KNOTWEED

CHICORY

DANDELION

COMMON
RAGWEED

123

These plants are important to know because they are toxic or irritating to people. Do not touch!

STINGING NETTLE   **To 4 ft.**

The Stinging Nettle, found in moist, shady places throughout North America, has stems covered with easily dislodged hollow hairs that are filled with formic acid, the same toxin that makes ant bites so painful. Look for *toothed* leaves opposite each other on the stem and *trailing ropes of tiny green flowers* growing from the sites where leaves join the *four-sided, hairy* stem.

POISON IVY   **Leaflets 2–4 in.**

Beware of any plant with *three leaflets* making up the leaf; this dangerous plant is otherwise quite variable. The leaves may be shiny or dull, lobed or not, and the plant may be vinelike or more upright. The toxic oil is found in all parts of the plant except the pollen. It can be carried on clothing, tools, pets, or in smoke if plants are burned. Washing with plenty of water soon after exposure can help.

POISON OAK   **Leaflets 3–7 in.**

This western relative of Poison Ivy is equally variable in form. It often appears as a shrub or small tree and has *larger, rounder* leaflets. Like Poison Ivy, it turns a beautiful shade of red in the fall. Don't pick it!

POISON SUMAC   **To 15 ft.**

Much rarer than Poison Ivy or Poison Oak, this swamp-loving plant has given a bad name to its harmless relatives, the handsome, red-fruited sumacs. Poison Sumac carries its flowers and *whitish* fruits *drooping down* along its stem, hidden among its leaflets. It is most common east of the Mississippi, especially in the Great Lakes region.

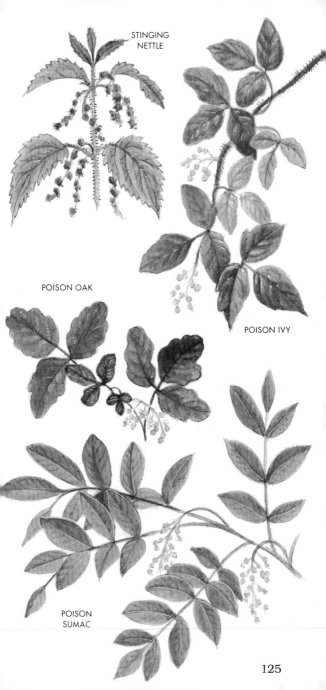

STINGING
NETTLE

POISON IVY

POISON OAK

POISON
SUMAC

125

TREE OF
HEAVEN

TREE OF HEAVEN   To 100 FT.

This common tree was brought to North
American cities because it thrives in poor
conditions, growing up to 8 feet per year.
Look for the telltale "thumb" at the base of
each leaflet and handsome clusters of
creamy or reddish seeds in late summer. If
you crush the leaves you will smell a faint
odor of old cigarette filters.

BLUEGUM
EUCALYPTUS

pods

BLUEGUM EUCALYPTUS   To 100 FT.

Originally from Australia, the eucalyptus
tree has long, pointed, leathery leaves, bark
that sheds in strips, and woody seed pods.
Its pungent oil, carried to the ground by "fog
drip," discourages the growth of other
plants. It reproduces naturally along the
coast of California.